Collins

AQA GCSE 9-1
Chemistry

T0328088

Sam Holyman

Acknowledgements

The authors and publisher are grateful to the copyright holders for permission to use quoted materials and images.

Every effort has been made to trace copyright holders and obtain their permission for the use of copyright material. The authors and publisher will gladly receive information enabling them to rectify any error or omission in subsequent editions. All facts are correct at time of going to press.

All images ©Shutterstock and HarperCollins*Publishers*

Published by Collins
An imprint of HarperCollins*Publishers* Limited
1 London Bridge Street
London SE1 9GF

HarperCollins*Publishers*
Macken House
39/40 Mayor Street Upper
Dublin 1, D01 C9W8, Ireland

© HarperCollins*Publishers* Limited 2024

ISBN 978-0-00-867231-7
First published 2024

10 9 8 7 6 5 4 3 2 1

British Library Cataloguing in Publication Data.

A CIP record of this book is available from the British Library.

Author: Sam Holyman
Publisher: Clare Souza
Commissioning: Richard Toms
Project Management and Editorial: Richard Toms and Katie Galloway
Inside Concept Design: Ian Wrigley
Layout: Rose & Thorn Creative Services Ltd
Cover Design: Sarah Duxbury
Production: Bethany Brohm

Printed in India by Multivista Global Pvt Ltd.

MIX
Paper | Supporting
responsible forestry
FSC
www.fsc.org **FSC™ C007454**

This book contains FSC™ certified paper and other controlled sources to ensure responsible forest management.

For more information visit: www.harpercollins.co.uk/green

How to use this book

Each topic is presented
on a two-page spread

Organise your
knowledge
with concise
explanations
and examples

Key points
highlight
fundamental
ideas

Higher tier
content is
highlighted
with a yellow
background
and the
HT logo

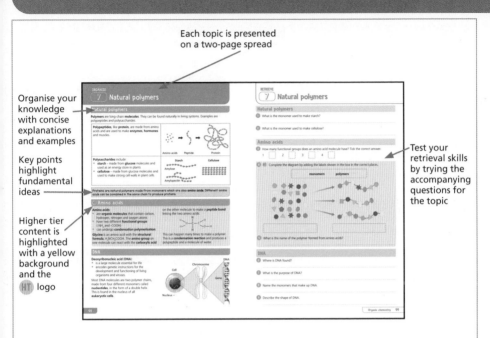

Test your
retrieval skills
by trying the
accompanying
questions for
the topic

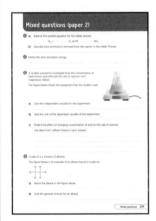

Mixed questions further test
retrieval skills after all topics
have been covered

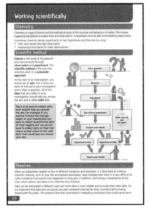

Scientific and maths skills
sections provide further
knowledge and explanations
of scientific and maths ideas
and investigative skills

Answers are provided to all
questions at the back of
the book

Contents

Contents

⑥ Paper 2: The rate and extent of chemical change

⑦ Paper 2: Organic chemistry

⑧ Paper 2: Chemical analysis

⑨ Paper 2: Chemistry of the atmosphere

⑩ Paper 2: Using resources

The difference between atoms, ions and isotopes

Atoms

Substances are made of particles. An **atom** is the smallest particle that can exist on its own. In an element, all the atoms are all the same and there are just under 100 naturally occurring elements. All elements are listed on the **periodic table**.

Atoms are made of **subatomic particles**. Atoms of the same element have the same number of **protons** in the nucleus.

Every atom can be represented with a nuclear symbol. This gives the chemical symbol of the element as well as information about the subatomic particles it contains.

Helium atoms   **Oxygen molecules**

Parts of an atom

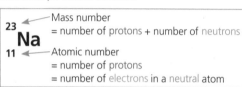

Mass number
= number of protons + number of neutrons

Atomic number
= number of protons
= number of electrons in a neutral atom

Electron
(negative electrical charge)

Proton
(positive electrical charge)

Neutron
(neutral)

Nucleus
(protons and neutrons)

Ions

Ions are atoms with a charge. Metal elements:
- are on the left and towards the bottom of the periodic table
- form positive ions by losing **electrons**.

Non-metal elements:
- are on the right and the top of the periodic table
- become negative ions by gaining electrons.

Neutral atom

Loss of
electron(s)

Gain of
electron(s)

Cation   **Anion**

Isotopes

Isotopes are different forms of an atom of an element. Isotopes of the same element have:
- the same **atomic number**
- a different **mass number**.

Carbon 12   **Carbon 13**   **Carbon 14**

6 protons   6 protons   6 protons
6 neutrons   7 neutrons   8 neutrons

Isotopes of the same element have the same number and arrangement of electrons. The electrons 'do' the chemistry and so they have the same chemical properties.

However, isotopes of the same element have a different mass number and so a different number of neutrons. This means that they have slightly different physical properties.

RETRIEVE 1 The difference between atoms, ions and isotopes

Atoms

1 What is an atom?

2 Why is hydrogen an element?

3 Approximately how many naturally occurring elements are there?

4 What are the names of the **three** subatomic particles?

Ions

5 What is an ion?

6 **a)** What charge do metal ions have?

b) What charge do non-metal ions have?

7 What sort of element is found on the left and towards the bottom of the periodic table?

Isotopes

8 What are isotopes?

9 **a)** How is the structure of isotopes of the same element similar?

b) How is the structure of isotopes of the same element different?

10 How many isotopes does carbon have?

The difference between compounds and mixtures

Compounds

Compounds are:
- more than one type of **atom** chemically joined
- formed from **elements** by **chemical reactions**.

Every compound has its own **formula**. A formula shows the symbols of the elements that it is made from and then subscript numbers to show how many of each type of atom is present.

Compounds can only be separated into their elements by chemical reactions.

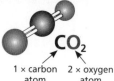

CO_2

1 × carbon atom   2 × oxygen atom

When the elements carbon and oxygen undergo a **chemical reaction (combustion)** the compound carbon dioxide is made.

Mixtures

Mixtures are made from more than one substance not chemically joined. In a mixture, the chemical properties of each substance are unchanged.

Some mixtures are designed to be useful products called **formulations**. They are usually made by mixing the components in carefully measured quantities so the product has the desired properties.

A mixture of two different gases

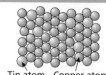

Tin atom   Copper atom

A metal alloy like bronze – an example of a formulation

Compounds melt and boil at specific temperatures but mixtures melt and boil over a range of temperatures. By measuring the melting point and boiling point of a substance, it is possible to classify a substance as a mixture.

Mixtures can be separated by a physical process as no new substances are made. Separation techniques include:

Chromatography
e.g. Inks and dyes. Coloured inks from a pen: The number of colours can be seen and the same ink identified from more than one sample.

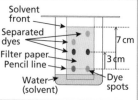

Solvent front
Separated dyes
Filter paper
Pencil line
Water (solvent)
7 cm
3 cm
Dye spots

Evaporation
e.g. Solutes from solvents in solutions. Sugar water: Small sugar crystals are formed and the solvent (water) is lost to the atmosphere.

Water vapor
Gauze mat
Tripod
Evaporating dish / beaker
Mixture
Bunsen burner

Filtering
e.g. Insoluble solid from a liquid. Sandy water: Sand is collected in the filter paper and the water in the test tube.

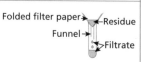

Folded filter paper
Funnel
Residue
Filtrate

Crystallisation
e.g. Solutes from solvents in solutions. Sugar water: Large sugar crystals form in a saturated sugar solution.

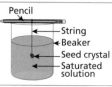

Pencil
String
Beaker
Seed crystal
Saturated solution

Distillation
e.g. Solvent from solutes in solutions. Tap water: Pure water is collected in the beaker and solutes are left in the round bottom flask.

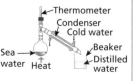

Thermometer
Condenser
Cold water
Beaker
Distilled water
Sea water   Heat

Fractional distillation
e.g. Separating a liquid mixture where each liquid has a different boiling point. Crude oil: As the temperature increases, each fraction (liquid part) of the mixture evaporates and the vapour then condenses into the receiving flask.

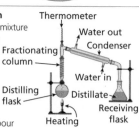

Thermometer
Water out
Condenser
Fractionating column
Water in
Distilling flask
Distillate
Receiving flask
Heating

RETRIEVE 1 — The difference between compounds and mixtures

Compounds

1 In what type of reaction are compounds made?

_____

2 a) How many atoms are in a molecule of carbon dioxide?   _____

b) How many different elements are in a molecule of carbon dioxide?   _____

3 In what type of reaction can compounds be separated back into their elements?

_____

Mixtures

4 What is a mixture?

_____

_____

5 What is a formulation?

_____

_____

6 Which **two** elements are in bronze?

_____

7 a) What sort of mixture can be separated by filtering?

_____

b) Which separation technique would be used to make pure water from sea water?

_____

c) How is distillation different to crystallisation? Give **one** similarity and **one** difference.

_____

_____

_____

_____

d) Which separation technique is used to separate crude oil into more useful mixtures?

_____

ORGANISE 1 — The development of the model of the atom

Scientific models

Scientific **models** are simplified representations of what is really happening. Good scientific models can be used to understand **observations** and make **predictions**.

Over time, scientific models change or get replaced as new **data** (evidence) is collected from using new technology and through further investigations.

Early atomic models

Atoms were initially thought to be solid spheres that could not be divided.

As technology developed in the 1800s, the **electron** subatomic particle was discovered by J.J. Thomson. He modified the atomic model to form the plum pudding model of the atom, which has:
- a ball of positive charge
- negative electrons embedded in it.

Solid sphere model (Dalton, 1803) **Plum pudding model (Thomson, 1897)**

Development of the model of the atom

Many scientists worked on trying to understand atoms better. In experiments by Geiger and Marsden, **alpha particles** (positive particles) were fired at thin sheets of gold foil. Most passed through, but some were deflected. They concluded that that the atom was mainly empty space. Their data was used by Rutherford, who concluded that there was a small positive **nucleus** in the centre of the atom.

Niels Bohr (and others) used maths to show that electrons **orbit** the nucleus at specific distances. His theory was tested and agreed with real-life experimental **observations**.

The last subatomic particle to be found was the **neutron**, discovered by James Chadwick. Being uncharged, it was hard to detect in experiments.

Nuclear model (Rutherford, 1911)

Planetary model (Bohr, 1913)

The Geiger and Marsden experiment

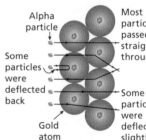

GCSE model of the atom

The GCSE model of an atom only works well for the first 20 elements.

Every atom has protons in the nucleus and electrons in the shells. Most atoms have neutrons in the nucleus. Hydrogen, $^1_1$H, is the only atom that doesn't have any neutrons.

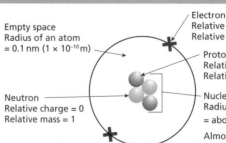

Empty space
Radius of an atom = 0.1 nm (1×10^{-10} m)

Neutron
Relative charge = 0
Relative mass = 1

Electron
Relative charge = −1
Relative mass = very small

Proton
Relative charge = +1
Relative mass = 1

Nucleus
Radius $< \frac{1}{10\,000}$ of the atom = about 1×10^{-14} m

Almost all of the mass of an atom is in the nucleus.

The development of the model of the atom

Scientific models

1 Why are scientific models useful?

....

....

2 Why do scientific models change?

Early atomic models

3 Before the electron was discovered, how were atoms described?

....

....

4 What was J.J. Thomson's model of the atom called?

....

Development of the model of the atom

5 What conclusion did Geiger and Marsden make from the results of their alpha scattering experiment?

....

6 What conclusion did Rutherford make from the results of the alpha scattering experiment?

....

....

7 Which scientist discovered the neutron?

....

GCSE model of the atom

8 Complete the sentences about the model of the atom by filling in the gaps.

Most of the mass of an atom is found in the ....

0.1 nm or 1×10^{-10} m is the size of the .... of an atom.

The subatomic particles found in the nucleus are .... and

....

The charge of a proton is .... and the charge of a neutron is

....

The mass of a neutron is ....

Determining the subatomic particles in an atom

Nuclear symbol

Every atom can be represented with a nuclear symbol. This gives the chemical symbol of the **element** as well as information about the **subatomic particles** it contains.

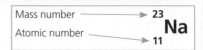

Mass number ⟶ 23
Atomic number ⟶ 11
Na

Calculating subatomic particles

For a neutral atom:
- the number of **protons** in the **nucleus** = **atomic (proton) number**
- the number of **electrons** in the shells = atomic number
- the number of **neutrons** in the nucleus = **mass number** – atomic number

Each atom of the same element has the same number of protons and so the same atomic number. But some elements have **isotopes**, which have a different number of neutrons in the nucleus.

We can use the nuclear symbol to calculate the number of subatomic particles in an atom of an element.

Element	Nuclear symbol	Mass number	Atomic number	Number of protons	Number of electrons	Number of neutrons
Hydrogen	$_{1}^{1}H$	1	1	1	1	1 – 1 = 0
Helium	$_{2}^{4}He$	4	2	2	2	4 – 2 = 2
Lithium	$_{3}^{7}Li$	7	3	3	3	7 – 3 = 4
Beryllium	$_{4}^{9}Be$	9	4	4	4	9 – 4 = 5
Boron	$_{5}^{11}B$	11	5	5	5	11 – 5 = 6
Carbon	$_{6}^{12}H$	12	6	6	6	12 – 6 = 6
Nitrogen	$_{7}^{14}N$	14	7	7	7	14 – 7 = 7
Oxygen	$_{8}^{16}O$	16	8	8	8	16 – 8 = 8
Fluorine	$_{9}^{19}F$	19	9	9	9	19 – 9 = 10
Neon	$_{10}^{20}Ne$	20	10	10	10	20 – 10 = 10

Using each symbol on the periodic table, you:
✓ **can** work out the number of protons in the nucleus (atomic (proton) number)
✓ **can** work out the number of electrons in the shells (atomic number)
✗ **cannot** calculate the number of neutrons in the nucleus.

The periodic table does **not** display nuclear symbols. It shows the **relative atomic mass**, not the atomic mass, of each element:

relative atomic mass
atomic symbol
name
atomic (proton) number

Relative atomic mass is the mass of an 'average' atom of that element, considering the amount of each isotope present in a typical sample. This means that the relative atomic mass is not always a whole number (e.g. $_{17}^{35.5}Cl$).

It is not possible to have half a subatomic particle. So, the symbols in the periodic table cannot be used to calculate the number of neutrons in an atom of that element.

> The atomic number is also called the proton number. It is the number of protons in an atom of an element. From this, you can infer that it is also the number of electrons in the neutral atom. In chemical reactions and physical changes, the atomic (proton) number of an atom or ion stays the same.

Determining the subatomic particles in an atom

1

Nuclear symbol

1 **a)** What does the upper number of the nuclear symbol represent?

....

b) What does the lower number of the nuclear symbol represent?

....

Calculating subatomic particles

2 **a)** What is the name of the element with the nuclear symbol $^{23}_{11}\text{Na}$?

....

b) How many protons are in an atom of $^{23}_{11}\text{Na}$?

....

c) How many electrons are in an atom of $^{23}_{11}\text{Na}$?

....

d) How many neutrons are in an atom of $^{23}_{11}\text{Na}$?

....

3 What information can the atomic (proton) number give you about the number of subatomic particles?

....

4 How do you calculate the number of neutrons in an atom?

....

5 What information can a symbol in the periodic table give you about the subatomic particles in the atoms of a particular element?

....

....

6 Why is the symbol of an element in the periodic table **not** a nuclear symbol?

....

....

7 Define atomic (proton) number.

....

....

....

The arrangement of electrons in the first 20 elements

Arrangement of electrons

Chemistry is the movement of **electrons** between atoms, ions or molecules. Electrons:

- fill each **energy level** (electron shell) in turn, starting with the one closest to the nucleus
- are kept in an **orbit** because of the **electrostatic attraction** between the negative electrons and the positive **nucleus**.

The electrons in the outer shell of an atom are responsible for its **chemical properties**.

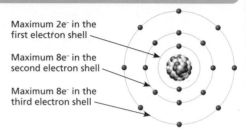

Maximum 2e⁻ in the first electron shell

Maximum 8e⁻ in the second electron shell

Maximum 8e⁻ in the third electron shell

Electronic structure of the first 20 elements

The **electronic structure** of these first elements can be shown in a diagram or by using digits (separated by commas, e.g. 2,1) to show how many electrons are in each energy level.

> The group number is the same as the number of outer shell electrons of the atoms in that column of the periodic table. The period number is the same as the number of occupied electron shells of atoms in that row of the periodic table.

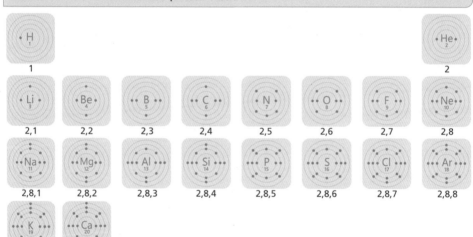

H 1							He 2
1							2
Li 3	Be 4	B 5	C 6	N 7	O 8	F 9	Ne 10
2,1	2,2	2,3	2,4	2,5	2,6	2,7	2,8
Na 11	Mg 12	Al 13	Si 14	P 15	S 16	Cl 17	Ar 18
2,8,1	2,8,2	2,8,3	2,8,4	2,8,5	2,8,6	2,8,7	2,8,8
K 19	Ca 20						
2,8,8,1	2,8,8,2						

Electronic structure of ions

Ions are charged atoms with a full outer shell of electrons. **Metal** ions have fewer electrons than the metal atom and **non-metal** ions have more electrons than the non-metal atom.

The electronic structure of an ion is shown within a pair of square brackets with the charge written outside at the top right corner.

Sodium ion
2,8

Fluoride ion
2,8

$^{23}_{11}Na^+$

11 protons

10 electrons (11 − 1)

12 neutrons

Sodium is a metal. It forms positive ions when it loses an electron.

$^{19}_{9}F^-$

9 protons

10 electrons (9 + 1)

10 neutrons

Fluorine is a non-metal. It forms negative ions when it gains an electron.

The arrangement of electrons in the first 20 elements

Arrangement of electrons

1 **a)** How many electrons complete the first energy level of an atom? ....

b) How many electrons complete the second energy level of an atom? ....

c) How many electrons complete the third energy level of an atom? ....

2 In what sequence do electrons fill energy levels in an atom?

....

....

Electronic structure of the first 20 elements

3 How many electrons do all Group 1 elements have in their outer shell? ....

4 How many electron shells are occupied in Period 3 elements? ....

5 Draw the electronic structure of a lithium atom.

6 Which element has atoms in the electronic structure 2,8,8? ....

Electronic structure of ions

7 What is an ion?

....

8 **a)** What is the charge on a sodium ion? ....

b) Draw the electronic structure of a sodium ion.

9 Explain why both sodium and fluoride ions have the same electronic structure.

....

....

....

....

The development of the periodic table

Classification

In science, objects with similar properties are **classified** (grouped) so that:
- they can be described easily without confusion
- connections can be seen between different objects
- **predictions** can be made about new objects that are found and fit into a particular group.

Scientists tried to organise **elements** into **groups** and classify them into tables before **subatomic particles** had been discovered. These tables were not very useful because:
- they were incomplete as not all elements had been discovered
- ordering was done using increasing atomic weight and some elements were placed in inappropriate groups.

Mendeleev's periodic table

In the 1800s, Dmitri Mendeleev overcame some of the problems of sorting the elements by:
- leaving gaps for elements still to be found
- broadly placing the elements in order of increasing atomic weight, but swapping some elements so that they were grouped by properties.

As technology improved, more elements were discovered and Mendeleev's predictions were shown to be correct. The discovery of subatomic particles and isotopes further supported the order of elements that Mendeleev developed.

The modern periodic table

There are some differences between Mendeleev's periodic table and the modern version:
- Information about the **atomic (proton) number** has been added for each element.
- **Atomic mass** is used rather than atomic weight.
- The four gaps left by Mendeelev have been filled

with the names, symbols and experimentally measured information about each element.
- An additional 50+ elements are included.
- An extra main column, Group 0 (**noble gases**), has been added.
- **Transition elements** are in a separate block.

Groups		Metals										3	4	5	6	7	0	Periods
1	2	Non-metals																
				H													He	1
Li	Be											B	C	N	O	F	Ne	2
Na	Mg											Al	Si	P	S	Cl	Ar	3
K	Ca	Sc	Ti	V	Cr	Mn	Fe	Co	Ni	Cu	Zn	Ga	Ge	As	Se	Br	Kr	4
Rb	Sr	Y	Zr	Nb	Mo	Tc	Ru	Rh	Pd	Ag	Cd	In	Sn	Sb	Te	I	Xe	5
Cs	Ba	La	Hf	Ta	W	Re	Os	Ir	Pt	Au	Hg	Tl	Pb	Bi	Po	At	Rn	6
Fr	Ra	Ac	Rf	Db	Sg	Bh	Hs	Mt	Ds	Rg	Cn	Nh	Fl	Mc	Lv	Ts	Og	7

The elements in the modern periodic table are arranged by increasing atomic (proton) number, and in groups (columns) where elements have similar properties.

Most elements are **metals** and solid at room temperature (20°C). The position of an element in the periodic table can be used to predict its physical and chemical properties.

> The properties of the elements repeat at regular intervals in the periodic table. Similar chemical properties are seen in elements of the same group because they have the same number of electrons in their outer electron shell.

① The development of the periodic table

Classification

1 What does classification mean?

_____

_____

2 Why do scientists classify objects?

_____

_____

_____

_____

3 Explain why early attempts to order the elements failed.

_____

_____

_____

Mendeleev's periodic table

4 **a)** Describe how Mendeleev ordered the elements in the periodic table.

_____

_____

_____

b) Why did Mendeleev leave gaps in his periodic table?

_____

_____

c) What new data justified Mendeleev's order of elements in the periodic table?

_____

_____

_____

The modern periodic table

5 How are the elements arranged in the modern periodic table?

_____

6 Which group of elements was added to the modern periodic table?

_____

7 Complete this sentence.

Most elements can be classified as _____.

(1) Group 1

Group 1 elements

Group 1 elements are **alkali metals**. They make up the first column of the periodic table. They:

- are metallic elements and so are **conductors**, **malleable**, **ductile** and **lustrous** (shiny)
- have **one electron** in the outer shell
- form **1+ ions** by losing the electron in their outer shell.

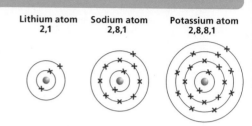

Lithium atom
2,1

Sodium atom
2,8,1

Potassium atom
2,8,8,1

Properties of Group 1 elements

The **melting points** and **boiling points** of Group 1 elements decrease as you go down the group because there is a decrease in the force of attraction between the atoms.

Their **densities** increase as you go down the group.

Since all Group 1 elements have one electron in their outer shell, they have similar chemical properties. As you go down Group 1:

- atoms get larger
- there is less attraction between the outer-shell electrons and the positive nucleus
- atoms more easily form a 1+ ion
- the reactivity of the elements increases.

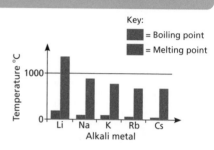

Key:
= Boiling point
= Melting point

General equations can be used to summarise the common reactions of alkali metals:

- **alkali metal + oxygen → metal oxide**
- **alkali metal + water → metal salt + hydrogen**
- **alkali metal + chlorine → metal chloride**

Symbol	Reaction with oxygen	Reaction with chlorine	Reaction with water
Li	$4Li(s) + O_2(g) \rightarrow 2Li_2O(s)$ In a Bunsen burner, it burns with a red flame. Open to the air, it slowly tarnishes.	$2Li(s) + Cl_2(g) \rightarrow 2LiCl(s)$ Burns slowly with a red flame and a white solid is produced.	$2Li(s) + 2H_2O(l) \rightarrow 2LiOH(aq) + H_2(g)$ Floats and moves on the surface of the water as it effervesces. Gets smaller and seems to disappear.
Na	$4Na(s) + O_2(g) \rightarrow 2Na_2O(s)$ In a Bunsen burner, it burns with a yellow flame. Open to the air, it tarnishes quickly.	$2Na(s) + Cl_2(g) \rightarrow 2NaCl(s)$ Burns quickly with a bright yellow flame and a white solid is produced.	$2Na(s) + 2H_2O(l) \rightarrow 2NaOH(aq) + H_2(g)$ Floats and moves fast on the surface of the water as it effervesces. Melts into a sphere, gets smaller and seems to disappear.
K	$4K(s) + O_2(g) \rightarrow 2K_2O(s)$ In a Bunsen burner, it burns with a lilac flame. Open to the air, it tarnishes very quickly.	$2K(s) + Cl_2(g) \rightarrow 2KCl(s)$ Burns very brightly with a lilac flame and a white solid is produced.	$2K(s) + 2H_2O(l) \rightarrow 2KOH(aq) + H_2(g)$ Floats and moves very fast on the surface of the water as it effervesces. Ignites into a lilac flame, gets smaller and seems to disappear.

In chemical reactions, the alkali metal atoms lose one electron to become positive metal ions. This is an example of an oxidation reaction and can be shown in a diagram and a half-equation:

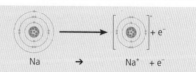

Na   →   Na⁺   + e⁻

Group 1 elements

1 Where are the Group 1 elements found in the periodic table?

....

2 What are the Group 1 elements also called?

....

3 Explain how the electron configuration is similar for all Group 1 metals.

....

4 What charge do Group 1 metal ions have? ....

Properties of Group 1 elements

5 **a)** What is the trend in the melting point as you go **down** Group 1?

....

b) Why does boiling point decrease as you go **down** Group 1?

....

....

c) What is the trend in density as you go **up** Group 1?

....

d) How does reactivity change as you go **down** Group 1?

....

6 What substance is made when rubidium reacts with oxygen?

....

7 What do you observe when sodium reacts with chlorine?

....

....

8 What colour flame is seen when lithium reacts with water? Tick the correct option.

Blue ☐

Turquoise ☐

Red ☐

Lilac ☐

9 Describe what is meant by effervescence.

....

....

1 Group 7

Group 7 elements

Group 7 **elements** are in the seventh main column of the **periodic table**. Group 7 elements:

- are called the **halogens**
- are non-metal elements and so are **dull, poor conductors** and **brittle**
- have **seven electrons** in the outer shell
- form **1– ions** by gaining an electron in the outer shell.

Fluorine atom
2,7

Chlorine atom
2,8,7

Properties of Group 7 elements

The Group 7 elements are made of **molecules** formed by a pair of halogen atoms, e.g:

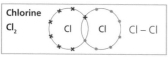

Chlorine
Cl_2 Cl   Cl   Cl – Cl

The **melting point** and **boiling point** increase as you go down the group because there is a stronger force of attraction between the molecules. So, at room temperature (20°C), fluorine and chlorine are gases, bromine is the only non-metal liquid and iodine is a solid. As you go down the group, the halogens become darker in **colour**.

There is an increase in **relative formula mass** of the halogen molecule as you go down the group because the atoms get larger and heavier (they contain more subatomic particles).

All Group 7 elements have seven electrons in their outer shell. As you go down Group 7:

- the atoms get larger
- there is less attraction between the electrons in the outer shell and the positive nucleus
- atoms find it more difficult to form 1– **halide** ions
- **reactivity** decreases.

Displacement reactions are when a more reactive halogen **displaces** a less reactive halogen from its **compound** (iodine (I_2) has no reaction).

Cl atom 2,8,7   **Cl⁻ ion [2,8,8]**

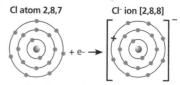

$+ e^- \rightarrow$

Chlorine, near the top of Group 7, is a highly reactive element that readily forms chloride (Cl⁻) ions

Halogen \ Halide salt	Chlorine Cl_2 (aq)	Bromine Br_2 (aq)	Iodine I_2 (aq)
Sodium chloride NaCl (aq)	No reaction	No reaction	No reaction
Sodium bromide NaBr (aq)	*Observation*: Solution darkens *Word equation*: chlorine + sodium bromide → sodium chloride + bromine *Balanced symbol equation*: Cl_2 (aq) + 2NaBr (aq) → 2NaCl (aq) + Br_2 (aq) *Balanced ionic equation*: Cl_2 (aq) + 2Br⁻ (aq) → 2Cl⁻ (aq) + Br_2 (aq)	No reaction	No reaction
Sodium iodide NaI (aq)	*Observation*: Solution darkens *Word equation*: chlorine + sodium iodide → sodium chloride + iodine *Balanced symbol equation*: Cl_2 (aq) + 2NaI (aq) → 2NaCl (aq) + I_2 (aq) Balanced ionic equation: Cl_2 (aq) + 2I⁻ (aq) → 2Cl⁻ (aq) + I_2 (aq)	*Observation*: Solution darkens *Word equation*: bromine + sodium iodide → sodium bromide + iodine *Balanced symbol equation*: Br_2 (aq) + 2NaI (aq) → 2NaBr (aq) + I_2 (aq) *Balanced ionic equation*: Br_2 (aq) + 2I⁻ (aq) → 2Br⁻ (aq) + I_2 (aq)	No reaction

1 Group 7

Group 7 elements

1 Where are the Group 7 elements found in the periodic table?

2 What are the Group 7 elements also called?

3 Explain how the electron configuration is similar for all Group 7 metals.

4 What charge do Group 7 non-metal ions have?

Properties of Group 7 elements

5 **a)** Describe the trend in the relative formula mass of the halogen molecule as you go down Group 7.

b) Why does boiling point increase as you go down Group 7?

c) What is the trend in the colour of the elements as you go down Group 7?

d) How does reactivity change as you go down Group 7?

6 What is a halide ion?

7 What is a halide salt?

8 Define the term displacement reaction.

9 Write a word equation for the displacement reaction between chlorine water and a solution of sodium bromide.

1 Group 0

Group 0 elements

Group 0 **elements** make up the last column of the **periodic table**:

The Group 0 elements:
- are called the **noble gases**
- are non-metal elements
- have a full outer shell of **electrons**
- are all gases at room temperature.

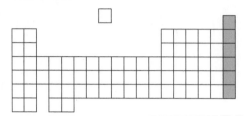

Helium atom	Neon atom	Argon atom
2	2,8	2,8,8

Properties of Group 0 elements

There is a gradual change in **physical properties** as you go down the group.

Physical properties include:
- **Boiling point** – increases as you go down the group because there is an increased force of attraction between the atoms.
- **Relative atomic mass** – increases as you go down the group.

All the elements in Group 0 have a stable arrangement of electrons with a full outer shell:
- Helium has two electrons in the outer shell.
- Other noble gases have 8 electrons in the outer shell.

This **electronic stability** means that the Group 0 elements are **inert** (unreactive). So, noble gases:
- do not easily form **molecules** – they are found as gaseous single atoms (**monatomic**) at room temperature (20°C)
- do not usually react with other **non-metals**
- do not easily form ions
- do not usually react with **metals**.

The correlation (pattern) between going down the noble gas group and the boiling point is more easily visualised with a bar chart.

Boiling point (°C) vs Noble gas

y-axis: 0, −25, −50, −75, −100, −125, −150, −175, −200, −225, −250, −273

x-axis: Helium, Neon, Argon, Krypton, Xenon

1 Group 0

Group 0 elements

1. Where are the Group 0 elements found on the periodic table?

2. What are the Group 0 elements also known as?

3. Explain how the electron configuration is similar for all Group 0 elements.

4. What state are all Group 0 elements found in at room temperature (20°C)?

Properties of Group 0 elements

5. **a)** What is the trend in the boiling point as you go down Group 0?

   b) What is the trend in relative atomic mass as you go down Group 0?

6. Describe the relationship between relative atomic mass and boiling point.

7. **a)** How many electrons are in the outer shell of helium?
   Tick the correct answer.

   0 ☐   2 ☐   6 ☐   8 ☐

   b) How many electrons are in the outer shell for all Group 0 elements except helium?
   Tick the correct answer.

   0 ☐   2 ☐   6 ☐   8 ☐

8. Why are Group 0 elements found as separate atoms?

9. Why do Group 0 elements rarely make compounds with:

   a) another non-metal?

   b) a metal?

1 Transition metals

Transition metal elements

Transition **elements** are in the central part of the **periodic table**, between Groups 2 and 3.

Transition metals are the metals that we often use in everyday life. Typical transition metals include chromium (Cr), manganese (Mn), iron (Fe), cobalt (Co), nickel (Ni) and copper (Cu).

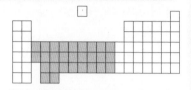

Properties of transition metal elements

The typical properties of transition elements are that they:
- are **shiny**, **ductile**, **malleable** and **conductors**
- make positive **ions** by losing their outer shell electron
- have more than one stable ion
- make coloured compounds.

When compared to Group 1 metals (alkali metals), transition metals have:
- higher **melting points** and **densities**
- greater strength and hardness.

As the transition metals have an **atomic (proton) number** of more than 20, you will not be expected to draw the electronic structure for these elements. The outer shell **electrons** are still involved in **chemical reactions**.

Transition metals:
- are less reactive than Group 1 metals – some do not react at all with oxygen, water or halogens, and others will react slowly
- have more than one stable ion
- form coloured compounds and solutions whereas Group 1 metal compounds are white and make colourless solutions.

> The physical properties are general properties of transition metals. There are always exceptions to the rule. For example, mercury (Hg) is a transition metal but is a liquid at room temperature (20°C) and so must have a low melting point, even lower than any group 1 element.

Copper and its ions

Copper reacts slowly with oxygen at room temperature (20°C) and the rate of this reaction can be increased by heating the metal in a Bunsen flame.

There is more than one type of stable copper ion, so there are two different types of copper oxide that can be made:
- copper(I) oxide, Cu_2O, made from copper(I) ions, Cu^+
- copper(II) oxide, CuO, made from copper(II) ions, Cu^{2+}

A Roman numeral is used in the metal compound name to show the positive charge on the metal ion.

So, the specific equations for the **oxidation** of copper are:
- copper + oxygen → copper(I) oxide
  $$2Cu(s) + O_2(g) \rightarrow 2CuO(s)$$
- copper + oxygen → copper(II) oxide
  $$4Cu(s) + O_2(g) \rightarrow 2Cu_2O(s)$$

Uses of transition metals

Transition metals can be used as **catalysts**. Catalysts include:
- iron used in the **Haber process** to make **ammonia**

- manganese(IV) oxide, Mn_2O, used to speed up the **decomposition** of hydrogen peroxide to oxygen and water
- nickel used to break a carbon-to-carbon double bond and add hydrogen.

1 Transition metals

Transition metal elements

1 Where are the transition elements found on the periodic table?

2 Give an example of a typical transition metal.

Properties of transition metal elements

3 Give **one** physical property that transition metals have in common with Group 1 metals.

4 **a)** How does the density of transition metals compare to Group 1 metals?

b) How does the melting point of transition metals compare to Group 1 metals?

c) How does the strength and hardness of transition metals compare to Group 1 metals?

d) How does the reactivity of transition metals compare to Group 1 metals?

e) How do the compounds of transition metals compare to the compounds of Group 1 metals?

Copper and its ions

5 **a)** Name the **two** stable copper ions.

b) Give the formulae of the **two** stable copper ions.

Uses of transition metals

6 What is the purpose of adding iron into the Haber process?

7 Name the catalysts that can be used to decompose hydrogen peroxide.

② Covalent bonding

Covalent bonding

A **covalent bond** is formed when two atoms share a pair of electrons.

All chemical bonds are strong. This means that a lot of energy is needed to break them. When a covalent bond is broken, the atoms are separated.

Non-metals use **covalent bonds** to:
• complete their outer **shell** of **electrons**
• have a stable **noble gas electron configuration**.

A chlorine molecule (one covalent bond)

2 chlorine atoms ⟶

A chlorine molecule ⟶ (made up of 2 chlorine atoms)

2 shared pair of electrons = covalent bond

Covalent bonding in elements and compounds

Other than the noble gases, most non-metal elements are found as molecules made up of two atoms. The table shows the **dot and cross diagrams** and **displayed formulae** for these **elements**.

	Chlorine Cl_2	Hydrogen H_2	Oxygen O_2	Nitrogen N_2
Dot and cross	Cl Cl	H H	O O	N N
Displayed formula	Cl – Cl	H – H	O = O (a double bond)	N ≡ N (a triple bond)

Many common substances make small molecules. The table shows the dot and cross diagrams and displayed formulae for these compounds.

Some elements, like diamond, make **giant structures** rather than molecules. Some compounds make giant structures like silicon dioxide, SiO_2.

	Water H_2O	Hydrogen chloride, HCl	Methane CH_4	Ammonia NH_3
Dot and cross	H O H	H Cl	H C H (with H top and bottom)	H N H (with H bottom)
Displayed formula	H–O–H	H–Cl	H–C–H (with H top and bottom)	H–N–H (with H bottom)

Polymers are giant structures made of very large molecules. They are made of repeating groups of atoms held together by covalent bonds.

Bottle tops can be made of polypropylene or PP.

Their structure is so long that we look for the small repeating unit and represent the whole structure by:
• writing the repeating structure in square brackets
• adding a lowercase 'n' to the outside bottom right of the bracket which means 'many of this unit'.

Representing structures

Scientific **models** can represent substances, but they have benefits and limitations:
• **Molecular formula** – only shows how many of which atom are present in the molecule.
• **Displayed formula** – shows every atom and every bond in the molecule but not shape or size of the molecule or the electrons.
• **Dot and cross diagram** – shows the electrons in relation to the atoms but does not show the shape or size of the molecule.

• **Ball and stick** – shows the number and type of atom present and suggests their size and position. But the atoms are too far apart from each other and the electrons are not visualised.
• **Diagrams** – 2D or 3D and show the number and arrangement of atoms, but no information about electrons. 3D computer models are useful as they rotate the structure.

2 Covalent bonding

Covalent bonding

1 What sort of elements can form a covalent bond?

....

2 What is a covalent bond?

....

Covalent bonding in elements and compounds

3 Which non-metal group of elements on the periodic table do not undergo covalent bonding?

....

4 Draw the dot and cross diagram for a chlorine molecule.

5 How many covalent bonds are there in a nitrogen molecule?

....

6 Draw the displayed formula for an ammonia molecule.

7 Define the term polymer.

....

....

Representing structures

8 What are the limitations of the ball and stick model to represent a covalent substance?

....

....

9 Which representation of covalent bonding shows the electrons? Tick the correct answer.

Dot and cross diagram ☐   Molecular formula ☐

Displayed formula ☐   Ball and stick model ☐

② Ionic bonding

Making metal ions and non-metal ions

Metal atoms lose all their outer shell **electrons** to get the **electronic structure of a noble gas**:
- Group 1 metals make 1+ ions, e.g. sodium, Na^+
- Group 2 metals make 2+ ions, e.g. magnesium, Mg^{2+}
- Group 3 metals make 3+ ions, e.g. aluminium, Al^{3+}
- Transition metals make 2^+ ions and at least one other stable ion.

Non-metal atoms gain electrons to get a stable noble gas electronic configuration:
- Group 5 metals make 3- ions, e.g. nitride, N^{3-}
- Group 6 metals make 2- ions, e.g. oxide, O^{2-}
- Group 7 metals make 1- ions, e.g. chloride, Cl^-

HT Ionic equations are more challenging for non-metal atoms as they usually start as **molecules**.

HT **Ionic equations** are balanced in terms of charge and particle. So, for making a magnesium ion:
- Dot and cross diagram:

Mg atom 2,8,2 → Mg^{2+} ion $[2,8]^{2+}$ + $2e^-$
- Ionic equation: $Mg \rightarrow Mg^{2+} + 2e^-$

For making an oxygen ion:
- Dot and cross diagram:

- Ionic equation: $O_2 + 4e^- \rightarrow 2O^{2-}$

Forming an ionic bond

Ionic bonds are made when a metal reacts with a non-metal. The outer shell electrons on the metal atom are donated. The outer shell of the non-metal ion accepts the electrons. The resulting **electrostatic force** of attraction between the positive and negative ions is the ionic bond.

Often, only the outer shell electrons are shown in **dot and cross diagrams**:

Ionic compounds

Ionic compounds form a giant structure known as a **lattice**, where the electrostatic forces of attraction between oppositely charged ions act in all directions. So, 3D models can be useful at showing the arrangement of ions.

In an ionic compound, the charge is balanced. So, there may be more of one ion than another, e.g. magnesium chloride, $MgCl_2$:

● Negatively charged chloride ions
● Positively charged sodium ions

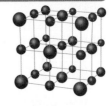

● Cl^- ● Na^+

Cl^- ion $[2,8,8]^-$ Mg^{2+} ion $[2,8]^{2+}$ Cl^- ion $[2,8,8]^-$

To work out the ionic compound formula
1. Write the formula of each ion next to each other, starting with the metal: Mg^{2+} Cl^{1-}
2. Circle the number of charges and cross them down:
   $Mg^{2+} \times Cl^{1-}$ becomes Mg_1Cl_2
3. Remove any 1s and cancel down: $MgCl_2$

The empirical formula is the smallest whole number ratio of atoms in a compound. The formula of any ionic compound is always the empirical formula of that compound.

2 Ionic bonding

Making metal ions and non-metal ions

1 How do metal atoms become ions?

....

2 What is the charge on a Group 2 metal ion?

....

3 Write the formula for a sodium ion.

....

4 How do non-metal atoms become ions?

....

....

5 What is the charge on a Group 7 non-metal ion?

....

6 Write the formula for an oxide ion.

....

Forming an ionic bond

7 **a)** Define the term ionic bond.

....

....

b) What types of elements can form an ionic bond?

....

c) How are electrons transferred to make an ionic bond?

....

Ionic compounds

8 What is the name of the structure formed by an ionic compound?

....

9 Give the formula for sodium oxide.

....

② Metallic bonding

Metallic bonding in pure metals

Pure metals are substances that contain atoms of only one metallic **element**. The atoms are held together in a **giant structure** by **metallic bonds**.

The structure of a pure metal has:
- a giant structure of atoms in a regular pattern
- free moving **delocalised electrons**.

Metallic bonds are made when the outer shell electrons from the metal atoms leave the atom. So:
- the outer shell electrons become able to move freely in the structure
- a regular pattern of positive metal **ions** is left behind.

The sharing of the delocalised electrons between the metallic ions in the giant structure is a metallic bond.

The bonding in metals can be represented by 2D diagrams.

> All pure metals are solids at room temperature, except for mercury, Hg, which is a liquid.

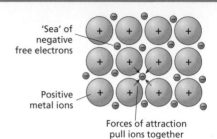

'Sea' of negative free electrons

Positive metal ions

Forces of attraction pull ions together

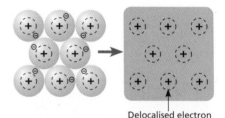

Delocalised electron

Properties of metals

The structure and bonding of a metal can be used to explain the general properties of metals:
- **Conductors** of electricity – delocalised electrons are free to move and carry the charge.
- **Ductile** and malleable (bends and shapes easily) – **planes** (layers) of atoms easily slide over each other.

- Conductors of heat – the delocalised electrons transfer the energy.
- High melting and boiling point – metallic bonds are strong and many bonds need to be broken to break the giant structure to melt or boil a metal; this requires a lot of energy.

Metallic bonding in alloys

Pure metals are usually quite soft and not used in everyday applications.

The addition of another element with a different sized atom distorts the arrangement of atoms, stopping them from being able to slide as easily. These are **alloys**; they are harder and have more useful properties for everyday uses.

Alloys can be described as:
- a **mixture** of a metal and at least one other element
- a **formulation** (a mixture designed for a useful product).

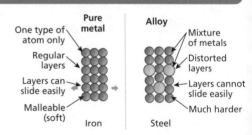

Pure metal

One type of atom only

Regular layers

Layers can slide easily

Malleable (soft)

Iron

Alloy

Mixture of metals

Distorted layers

Layers cannot slide easily

Much harder

Steel

> Metallic bonding can only occur in pure metals (metal elements) and alloys.

2 Metallic bonding

Metallic bonding in pure metals

1 Define what pure metals are.

....

....

2 **a)** What type of bonding is found in pure iron?

....

b) What sort of structure is found in pure iron?

....

c) How are the atoms arranged in a pure metal?

....

3 Explain what is meant by a metallic bond.

....

....

Properties of metals

4 Draw lines to match the property on the left to the correct explanation on the right.

Property	Explanation
Conductor of heat	Many strong metallic bonds must be broken.
Malleable (bends and shapes easily)	Delocalised electrons are free to move and carry charge.
High melting and boiling point	The layers of atoms easily slide over each other.
Conductor of electricity	Delocalised electrons can transfer the energy.

Metallic bonding in alloys

5 What is an alloy?

....

....

6 Name an alloy made mainly from iron.

....

② Giant structures and small molecules

Giant structures and their properties

Giant structures are 3D structures of **atoms** or **ions** held together by strong **bonds** so they have high melting points and a lot of energy is needed to break the bonds. There are three types:

Giant covalent structures	Giant ionic structures (lattices)	Giant metallic structures
Atoms are held by shared pair of electrons.	Ions are held in position by the **electrostatic** force of attraction between the oppositely charged ions. Lattices are often soluble in water and can **conduct** electricity when the ions are free to move and carry the charge, so as a liquid or in **aqueous solution**.	Found only in **pure metal elements** or **alloys**. Metal ions are held in a regular arrangement by the attraction of **delocalised** electrons.
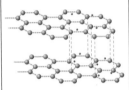 Graphite is made of hexagonal rings that stack as layers	 Positively charged ion Negatively charged ion The structure of an ionic compound is an ionic lattice	Close-packed atom

Small molecules and their properties

Molecules are small groups of **non-metal** atoms held together by covalent bonds. Many molecules are in a sample of a substance and there are weak forces of attraction between the molecules.

Strong covalent bond within the molecule   Weak forces of attraction between molecules

Simple molecules:
- are electrical **insulators** – no charged particles (ions or electrons) are free to move and carry a charge
- are soft and **brittle** – as the weak intermolecular forces of attraction are easily broken
- have low melting and boiling points – as only the weak forces of attraction between the molecules need to be overcome to melt or boil the substance. No strong covalent bonds are broken.

The melting point is always lower than the boiling point as a substance must melt before it can boil. As simple molecules get larger, the intermolecular forces of attraction also increase and so the melting and boiling points will be higher.

This relationship is easy to see in **homologous series** (families of chemicals). Each successive **alkane** molecule has an extra $-CH_2-$ group compared to the last, and is heavier and larger. This increases intermolecular forces of attraction and so more energy is needed to overcome them, leading to higher melting and boiling points.

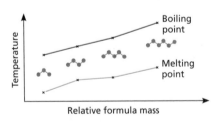

Group 0 elements have a stable outer shell of electrons. They have no bonds but do have weak forces of attraction between their atoms. As the atoms get larger, the weak forces of attraction increase. It is these weak forces of attraction that are overcome to melt and boil Group 0 elements. As these weak forces are very small, only a very small amount of energy is needed to overcome them and this is why these elements are found as gases at room temperature.

RETRIEVE

2 Giant structures and small molecules

Giant structures and their properties

1 What holds the atoms together in a giant covalent structure?

....

2 What is the name given to giant ionic structures?

....

3 What sort of substances can make giant structures held by metallic bonds?

....

4 In which states can ionic compounds conduct electricity? Tick the correct answers.

Solid [ ]   Gas [ ]

Liquid [ ]   Aqueous solution [ ]

5 Explain why giant structures have high melting and boiling points.

....

....

....

Small molecules and their properties

6 What sort of atoms make molecules?

....

7 How does the size of a molecule affect the melting and boiling points of a substance?

....

....

8 Draw lines to match the property on the left to the correct explanation on the right.

Property	Explanation
	Only relatively weak intermolecular forces of attraction need to be overcome and no bonds are broken.
Electrical insulator	
Low melting and boiling points	Weak intermolecular forces of attraction are easily broken.
Soft and brittle	No charged particles are free to move and carry a charge.

2) States of matter

States of matter

There are three states of matter, each with properties that can be explained by the particle model.

Each **atom**, or particle, of a substance has certain properties. These properties may not be the same as the **bulk properties**, which describe how a sample of the substance behaves.

The **particle model** allows us to explain observations about matter and make predictions about how matter will behave if conditions change. But there are limitations.

HT
- There are no forces between the particles, whereas in real life there are bonds between the particles in giant structures, and intermolecular forces of attraction between molecules.
- All particles are the same shape (spheres). Although atoms tend to be spheres, molecules are a wide variety of shapes.
- All particles are solid, though some molecules are cages, e.g. buckminsterfullerene.

Solids
- Particles are touching and arranged in an order.
- Each particle vibrates around a fixed position.
- Fixed shape and volume.

Liquids
- Particles are still touching but they are in a random arrangement.
- As the particles can move past each other, liquids can flow.
- No definite shape but a fixed volume.

Gases
- Particles move randomly in all directions with different speeds.
- Gases spread out to fill the container (diffuse) and can be poured.
- No definite shape or volume.

Changing state

A substance can change state when energy is added or taken away. As no new substance is made, this is an example of a **physical change**.

Melting point and **boiling point** can predict the state at different temperatures.

Each substance has a different strength of attraction between its particles so each needs a different amount of energy to change state when melting and boiling.

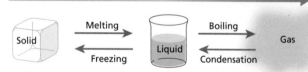

Energy transferred to the particles from the surroundings by heat

Solid — Melting → Liquid — Boiling → Gas
Solid ← Freezing — Liquid ← Condensation — Gas

Energy transferred from the particles to the surroundings by heat

Substance	Melting point (°C)	Boiling point (°C)	State at room temperature
Bromine (Br$_2$)	-7	59	Liquid
Oxygen (O$_2$)	-219	-183	Gas
Water (H$_2$O)	0	100	Liquid
Sodium chloride (NaCl)	801	1465	Solid

State symbols

State symbols are shown after the formula of the substance in an equation, for example:
- Solid, e.g. ice = H$_2$O(s)
- Liquid, e.g. water = H$_2$O(l)
- Gas, e.g. steam = H$_2$O(g)
- Aqueous solution (a solution where water is a solvent), e.g. brine (sodium chloride solution) = NaCl(aq).

② States of matter

States of matter

1 Complete the sentences about states of matter by filling in the missing words.

When a substance is a .... or a .... the

particles are able to move past each other.

When a substance is a .... the particles are in an ordered pattern.

2 Explain why solids cannot be poured.

....

....

3 Which states of matter have a fixed volume?

....

4 Explain what the particle model is used for.

....

....

....

5 Give an example of a particle that is a cage and so not a solid sphere.

....

Changing state

6 Why is changing state described as a physical change?

....

7 Name the change of state when a substance changes from a solid to a liquid.

....

8 What happens when a substance condenses?

....

....

9 Ammonia has a melting point of -77°C and a boiling point of -33°C. What is the state of ammonia at room temperature?

....

State symbols

10 What does (aq) mean?

....

11 What is the state symbol that you would use to show water at room temperature? ....

Carbon

Carbon is a **non-metal element** with the **electronic structure** 2,4. It is found in Group 4, Period 2 of the periodic table. It does not form any stable **ions** and so only makes **covalent bonds** in the elemental form with itself or with other non-metal **atoms** to form **compounds**.

> Carbon has many uses, but not every type of carbon is useful for every job.

Diamond, graphite and graphene

Diamond is a **giant covalent** structure. Each carbon atom is held in place by four strong covalent bonds.

- High melting and boiling points – many strong covalent bonds must be broken to pull apart the atoms.
- Electrical insulator (doesn't conduct electricity) – no charged particles (ions or **electrons**) that are free to move and carry the charge.
- Hard – the atoms are held in a ridged network.
- Used in drill bits – it is so hard and has such a high melting point that it can drill holes without being damaged itself.

Covalent bond between two carbon atoms

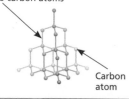

Carbon atom

Graphite is a giant covalent structure. Each carbon atom is held in hexagonal layers by three strong covalent bonds. Each carbon atom donates one outer shell electron to form **delocalised electrons**, which move freely through the structure. There are weak forces of attraction between the layers of carbon atoms.

- High melting and boiling points – many strong covalent bonds must be broken to pull apart the atoms.
- Electrical conductor – delocalised electrons move freely and carry charge.
- Soft – the layers (**planes**) of atoms easily slide over each other.
- Used in pencils as the layers of carbon atoms easily slide off and leave a mark on the paper.

Covalent bond between two carbon atoms

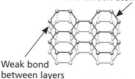

Weak bond between layers

Graphene is one layer of graphite. It is useful for making electronics and **composites**.

- High melting and boiling points – many strong covalent bonds must be broken to pull apart the atoms from the giant covalent structure.
- Electrical conductor – delocalised electrons are free to move and carry the charge.
- Flexible and strong – as the layers (planes) of atoms easily fold and bend while the strong covalent bonds do not break the atoms apart.
- Being developed into touch screens as highly flexible and an excellent electrical conductor.

Carbon atom

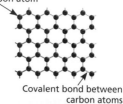

Covalent bond between carbon atoms

Fullerenes

Fullerenes are made only of carbon atoms and have a hollow shape. They are hexagonal rings of carbon atoms, but may also contain rings with five or seven carbon atoms. **Buckminsterfullerene** (C_{60}) was the first fullerene to be discovered and has a spherical shape.

Cylindrical fullerenes are called **carbon nanotubes**. They have very high length to diameter ratios and properties that make them useful for **nanotechnology**, electronics and materials.

Buckminsterfullerene

Carbon atom

Strong covalent bond

Nanotube

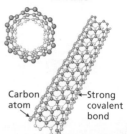

Carbon atom

Strong covalent bond

② The different structures of carbon

Carbon

1 How many electrons does carbon have in its outer shell?

....

2 a) What sort of bonding is found in a pure sample of carbon?

....

b) What sort of bonding is found in a compound containing carbon?

....

Diamond, graphite and graphene

3 a) How many strong bonds does each carbon atom have in diamond?

....

b) Why does diamond have very high melting and boiling points?

....

....

4 a) Where do the delocalised electrons come from in graphite?

....

b) How many strong bonds does each carbon atom have in graphite?

....

c) Why is graphite soft?

....

....

5 a) What is graphene?

....

....

b) Why can graphene conduct electricity?

....

....

Fullerenes

6 What is the formula of the first fullerene that was discovered?

....

7 What is **one** property of nanotubes?

....

Particles and their sizes

Substances can form clumps, called **particles**. The size of the particles is used to classify the substances:

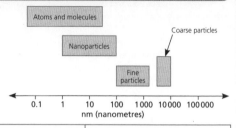

Atoms and molecules

Nanoparticles

Coarse particles

Fine particles

| 0.1 | 1 | 10 | 100 | 1000 | 10000 | 100000 |
nm (nanometres)

Dust
- Made of **coarse** particles.
- PM_{10}
- Diameter between $1 \times 10^{-5}\,m$ and $2.5 \times 10^{-6}\,m$.

Human hair

0.00001 metre $= 1 \times 10^{-5}\,m$

Can be seen using a microscope

Fine particles
- $PM_{2.5}$
- Diameter between 100 and 2500 nm.

Virus / small bacteria

0.0000001 metre $= 1 \times 10^{-7}\,m$

Can be seen using an electron microscope

Nanoparticles
- Made of a few hundred **atoms**.
- Between 1–100 nm in size.

Virus / small bacteria

0.000000001 metre $= 1 \times 10^{-9}\,m$

Nanoparticle zone

Properties and uses of nanoparticles

Nanoscience is the study and use of structures that are 1–100 nm in size.

As these structures have a high **surface to volume ratio**, a smaller **mass** is needed for them to be effective compared to normal particle sizes.

The smaller the particle, the greater the surface to volume ratio:
- As the side of cube decreases by a factor of 10, the surface area to volume ratio increases by a factor of 10.
- More atoms are exposed at the surface at the same time.
- This change means that the **bulk properties** of a large sample of the material are different to the properties of a nanoparticle of the same material.

Cut into eight equal-sized cubes

2 cm × 2 cm × 2 cm cube

1 cm × 1 cm × 1 cm each cube

The volume of a cube is calculated by multiplying the width by the length by the height of the cube. The units are cubed.

Modern tennis racquets use nanotechnology as carbon nanotubes are added to the frames to increase the strength, stability and power when hitting a tennis ball. **Carbon nanotubes** have a diameter of about 5 nm and are an example of nanotechnology.

Other uses of nanoparticles include medicine, electronics, cosmetics, sun creams, deodorants and **catalysts**.

Advantages of using nanoparticles	Disadvantages of using nanoparticles
• More sustainable as less mass of substance is needed for the same job. • Can penetrate into the human body, so medicines and sun creams are more easily absorbed.	• More expensive to make. • Effect in the body from absorption is not fully understood, so is a potential health risk. • Possible risk of polluting the environment: nanoparticles could build up in organisms.

② Nanoparticles

Particles and their sizes

1 Draw lines to match the particle on the left to the size on the right.

Particle	Size
Nanoparticle	Diameter between 1×10^{-5} m and 2.5×10^{-6} m
Fine particle	Between 1 and 100 nm in size
Dust	Diameter between 100 and 2500 nm

2 What is the classification of a $PM_{2.5}$ particle?

....

3 How many atoms are there in a nanoparticle?

....

Properties and uses of nanoparticles

4 How would you describe the surface to volume ratio of a nanoparticle?

....

5 Calculate the volume of a cube where each edge has a length of 2 nm.

....

6 What happens to the surface to volume ratio when the side of a cube decreases by a factor of 10?

....

7 Give **one** application of nanotechnology.

....

8 Why is nanotechnology especially useful for sun creams and medicines?

....

....

....

9 Why are some people worried about the impact on the environment of the use of nanotechnology?

....

....

....

....

ORGANISE 3 Relative formula mass and percentage composition

Relative atomic mass of elements

The **relative atomic mass** of an **element** is the weighted average mass of an atom of that element, considering the natural abundance of **isotopes**. It has the symbol A_r and is found on the **periodic table** for every element.

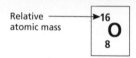

Relative formula mass

The **relative formula mass** of a **molecule** for a **non-metal** element is the sum of the atomic masses for each atom in the molecule, e.g. oxygen molecules (O_2) are made of two oxygen atoms **covalently** bonded together. So, the relative formula mass for a molecule of oxygen = $2 \times 16 = 32$.

The relative formula mass of a **simple covalent molecule** is the sum of the atomic masses for each atom in the molecule, e.g. water molecules (H_2O) are made of two hydrogen atoms covalently bonded to an oxygen atom.

Calculate the M_r of water, H_2O.

Write the formula	H_2O
Substitute the A_r values	$(2 \times 1) + 16$
Calculate the M_r	$2 + 16 = 18$

The empirical formula is the smallest whole number ratio of atoms in a compound.

The relative formula mass of an **ionic compound** is the sum of the atomic masses for the **empirical formula** of that compound, e.g.

Calculate the M_r of potassium carbonate, K_2CO_3.

Write the formula	K_2CO_3
Substitute the A_r values	$(39 \times 2) + 12 + (16 \times 3)$
Calculate the M_r	$78 + 12 + 48 = 138$

Sometimes the ratio of positive to negative ions is not 1:1. This is when brackets are used in formulae.

Calculate the relative formula mass of calcium nitrate, $Ca(NO_3)_2$.	Remember that everything inside a set of brackets is multiplied by the number outside the brackets, so $Ca(NO_3)_2$ contains 1 calcium, 2 nitrogen and 6 oxygen atoms.
Relative formula mass $= 40 + (14 \times 2) + (16 \times 6)$ $= 164$	

Percentage composition

- Percentage is a mathematical comparison of how much you have of something compared to the whole, where the whole is 100.
- So, **percentage composition** =

$$\frac{\text{mass of atoms of a certain element in a compound}}{\text{relative formula mass of the compound}} \times 100$$

The formulae of all ionic compounds are empirical formulae as this is the unit that is repeated many times in the ionic lattice.

Consider the percentage composition of carbon in carbon dioxide:

What is the relative formula mass of carbon dioxide, CO_2?

Relative formula mass $= (12 \times 1) + (16 \times 2)$ $= 44$

CO_2 contains 1 carbon atom with a relative atomic mass of 12, and 2 oxygen atoms with a relative atomic mass of 16.

So, the percentage composition of carbon in carbon dioxide $= \frac{12}{44} \times 100 = 27.3\%$

Relative formula mass and percentage composition

Relative atomic mass of elements

1 Where can you find the relative atomic masses of all the elements?

....

2 What is the relative atomic mass of carbon?

....

Relative formula mass

3 How do you calculate the relative formula mass for a molecule of an element?

....

....

4 **a)** What is the relative formula mass of hydrogen, H_2?

....

b) What is the relative formula mass of ozone, O_3?

....

5 How do you calculate the relative formula mass of a simple covalent molecule?

....

....

6 **a)** What is the relative formula mass of hydrogen peroxide, H_2O_2?

....

b) What is the relative formula mass of potassium oxide, K_2O?

....

7 How do you calculate the relative formula mass of an ionic compound?

....

....

Percentage composition

8 **a)** What is the percentage composition of oxygen in carbon dioxide, CO_2?

....

b) What is the percentage composition of hydrogen in water, H_2O?

....

c) What is the percentage composition of oxygen in water, H_2O?

....

③ Amount of substance

⊞ The mole

The **mole** is a measure of how many particles there are in a substance. The mole:

- is the chemical measure of the amount of substance
- is measured in **mol**
- has 6.02×10^{23} particles in it.

> In one mole of a substance there is the same number of particles per mole. As this number is always constant, it is called the **Avogadro Constant**.

For a mole of any element, the mass in grams is the same value as the relative atomic mass.

If you have one mole of a substance, the mass in grams is numerically equal to its relative formula mass. Consider the mass of one mole of sodium hydroxide, NaOH:

> A_r sodium + A_r hydrogen + A_r oxygen
>
> $= 23 + 1 + 16 = \textbf{40g}$

The formula connecting amount of substance, mass and relative formula mass is:

> $$\text{Number of moles of substance (mol)} = \frac{\text{mass of substance (g)}}{\text{mass of one mole (g/mol)}}$$

Example 1

Calculate the number of moles of carbon in 36g of the element.

$$= \frac{36g}{12 g/mol} = \textbf{3 moles} \quad \boxed{A_r \text{ carbon} = 12}$$

Example 2

Calculate the mass of 4 moles of sodium hydroxide. Rearrange the formula...

> $$\text{Mass of substance (g)} = \text{number of moles of substance (mol)} \times \text{mass of one mole (g/mol)}$$

$$= 4 \text{mol} \times 40 g/mol = \textbf{160g}$$

⊞ Amount of substance in equations

Balanced symbol equations work on ratios. This can be considered in terms of moles, for example:

- Oxidation of magnesium:
  - magnesium + oxygen → magnesium oxide
  - $2Mg(s) + O_2(g) \rightarrow 2MgO(s)$

2 moles of Mg react with 1 mole of O_2 to produce 2 moles of MgO.

Work out how much MgO is made. We know that the relative formula mass of:
Mg is 24; O_2 is 16×2 (= 32); and MgO is $24 + 16$ (= 40).

So if $2 \times 24g$ of Mg reacts with 32 g of O_2 then $2 \times 40g$ of MgO is made.

$2Mg + O_2 \rightarrow 2MgO = 48g + 32g \rightarrow 80g$

If 4.8g of Mg is used then 8.0g of MgO is made.

⊞ Using moles to balance equations

In the oxidation of magnesium:

> 72 g of magnesium was reacted with 48 g of oxygen molecules to produce 120 g of magnesium oxide. Use the number of moles of reactants and products to write a balanced equation for the reaction.

amount of Mg $= \frac{72}{24} = 3 \text{ mol}$	Use the masses of the reactants to calculate the number of moles present.
amount of $O_2 = \frac{48}{32} = 1.5 \text{ mol}$	Divide the number of moles of each substance by the smallest number (1.5) to give the simplest whole number ratio.
amount of MgO $= \frac{120}{40} = 3 \text{ mol}$	This shows that 2 moles of magnesium react with 1 mole of oxygen molecules to produce 2 moles of magnesium oxide.

$$3Mg + 1.5O_2 \rightarrow 3MgO$$

$$2Mg + O_2 \rightarrow 2MgO$$

Often the expensive or most hazardous reactant is carefully measured to make sure that the limiting reactant can fully react and the maximum amount of product is made.

> In a chemical reaction, the reactant that is fully used up is called the **limiting reactant** and the reactant(s) that is not fully used up is described as being in **excess**.

Amount of substance

HT The mole

1 Draw lines to match the key term on the left to the definition on the right.

| Mole | | Number of particles in a mole, 6.02×10^{23} |

| Molar mass | | The mass of one mole of a substance, which is equal to the M, value in grams |

| Avogadro constant | | The chemical measure of the amount of substance |

2 Write the formula that connects amount of substance, mass and relative atomic mass.

3 How many moles of oxygen atoms are in 32 g of oxygen molecules?

4 Write the formula that connects amount of substance, mass and relative formula mass.

HT Amount of substance in equations

5 How many moles of magnesium react with one mole of oxygen atoms in the oxidation of magnesium?

6 What mass of magnesium oxide is made when two moles of magnesium are completely oxidised?

HT Using moles to balance equations

7 What is a limiting reactant?

8 How do you ensure that a reactant is in excess?

③ Concentration

Concentration of solutions

Many chemical reactions take place in **solutions**. Solutions are a **mixture** made of a **solute** (usually a solid) **dissolved** into a **solvent** (usually a liquid).

We can use the **particle model** to imagine the process of dissolving:

- The solute particles fit into the gaps in between the solvent particles.
- The **mass** of the solution
  = mass of solvent + mass of solute.
- The **volume** of the solution is usually the same volume as the solvent.

Concentration is a measure of how much solute is dissolved in 1 dm³ of a solvent.

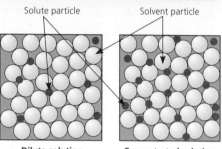

Solute particle   Solvent particle

Dilute solution
Not much solute in a
given volume of solvent

Concentrated solution
Lots of solute in a given
volume of solvent

Calculating concentration with mass

The concentration of a solution can be measured in mass per given volume of solution. Suitable **units** are grams per dm³, g/dm³ and gdm⁻³.

This **formula** can be used to work out the concentration of a solution:

$$\text{Concentration (g/dm}^3) = \frac{\text{mass (g)}}{\text{volume of solvent (dm}^3)}$$

2.00 dm³ of sodium hydroxide (NaOH) solution contains 20 g of sodium hydroxide.

Work out the concentration.
Concentration (g/dm³) =
20 g ÷ 2.00 dm³ =
10 g/dm³

Mass
(g)

Conc.   Volume
(g/dm³)   (dm³)

ⒽⓉ Calculating concentration with moles

The concentration of a solution can be measured in **amount of substance** per given volume of solution. Suitable units are moles per dm³, mol/dm³ and mol/dm⁻³.

$$\text{Concentration (g/dm}^3) = \frac{\text{amount of substance (mol)}}{\text{volume of solvent (dm}^3)}$$

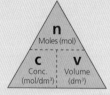

n
Moles (mol)

c   V
Conc.   Volume
(mol/dm³) (dm³)

2.00 dm³ of sodium hydroxide solution contains 0.50 moles of sodium hydroxide.

Work out the concentration of the solution.

Concentration of a solution	=	$\dfrac{\text{amount of substance (mol)}}{\text{volume (dm}^3)}$	Substitute the values
	=	$\dfrac{0.50\,\text{mol}}{2.00\,\text{dm}^3} = 0.25\,\text{mol/dm}^3$	into the formula.

It is possible to compare the concentration of a solution, but it is useful to do so in the same units:

- 1 dm³ of a 0.25 mol/dm³ sodium hydroxide solution would have 0.25 moles of sodium hydroxide per 1 dm³ of solution.
- The relative formula mass, M_r, of sodium hydroxide is 40.
- Using $n = m \div M_r$, we can calculate that there is a mass of 10 g of sodium hydroxide in this solution.
- Therefore, the 0.25 mol/dm³ concentration of sodium hydroxide solution is the same as 10 g/dm³.

1 dm³ = 1000 cm³ = 1 litre

To change cm³ into dm³, divide by 1000.

③ Concentration

Concentration of solutions

1 Draw lines to match the key term to the definition on the right.

Solute		The mixture

Solution		The substance that dissolves

Solvent		The substance that something dissolves in

2 How would you describe a concentrated solution?

....

....

3 How would you describe a dilute solution?

....

....

Calculating concentration with mass

4 Write down the formula that connects concentration, mass and volume.

....

....

5 What is the concentration of a solution made from 10 g of sodium hydroxide, NaOH, in 2.5 dm³ solution?

....

HT Calculating concentration with moles

6 Write down the formula that connects concentration, amount of substance and volume.

....

....

7 What is the concentration of a solution made from 0.5 moles of sodium hydroxide, NaOH, in 2.5 dm³ solution?

....

8 What are the **two** units that concentration can be measured in?

....

3 Yield

Yield and percentage yield

Yield is a measure of how much desirable **product** has been made and collected. In a chemical reaction, **mass** is conserved, as no atoms are gained or lost. A **top pan balance** can measure the mass of the desirable product collected. Not every atom in the **reactants** will become a desirable product. This is due to one or more of these reasons:

- the reaction is **reversible**
- during each transfer, some product is lost
- the separation of the product from the reaction mixture leaves some product behind
- other reactions take place.

The **percentage yield** is a comparison of how much product is collected (**actual yield**) compared to the **theoretical maximum** mass of product that could be made based on the limiting reactant.

Percentage yield (%)

$$= \frac{\text{mass of product actually made}}{\text{maxiumum theoretical mass of product}} \times 100$$

In reality, it is impossible to get more than 100% yield. But, in practice, it is possible if the product that is collected is not pure. This means there was an error in measuring the actual yield and a greater mass was recorded.

Example: 50 kg of calcium carbonate ($CaCO_3$) is expected to produce 28 kg of calcium oxide (CaO). A company heats 50 kg of calcium carbonate in a kiln and obtains 22 kg of calcium oxide.

Calculate the percentage yield.

Percentage yield $= \frac{22}{28} \times 100 = 78.6\%$

HT You need to be able to calculate the maximum theoretical mass that can be made before you can calculate the percentage mass of the product.

Calculate how much calcium oxide can be produced from 50.0 kg of calcium carbonate.

$$CaCO_3 \rightarrow CaO + CO_2$$

$$[40 + 12 + (3 \times 16)] \rightarrow [40 + 16] + [12 + (2 \times 16)]$$

$$100 \rightarrow 56 + 44$$

$$100 : 56$$

100 kg of $CaCO_3$ produces $\frac{56}{100} = 0.56$ kg of CaO

And, 50 kg of $CaCO_3$ produces $0.56 \times 50 = 28$ kg of CaO

Relative atomic masses: Ca = 40, C = 12, O = 16

Write down the equation and work out the M_r of each substance.

Check that the total mass of reactants equals the total mass of products.

The question only mentions calcium oxide and calcium carbonate, so ignore CO_2. You just need the ratio of mass of reactant to mass of product.

Use the ratio to calculate how much calcium oxide can be produced.

Atom economy

Atom economy (also called atom utilisation) is:

- a measure of the amount of starting materials that end up as useful products.

$$\text{Atom economy} = \frac{\begin{array}{c}\text{relative formula mass} \\ \text{of desired product} \\ \text{from the equation}\end{array}}{\begin{array}{c}\text{sum of all relative formula} \\ \text{masses of all reactants} \\ \text{from the equation}\end{array}} \times 100$$

Example

calcium carbonate $\rightarrow$ calcium oxide + carbon dioxide

The products of this reaction are calcium oxide (useful) and carbon dioxide (waste).

Calculate the atom economy of this reaction.

Calculate the M_r of the reactants and products.

$$CaCO_3 \rightarrow CaO + CO_2$$

M_r of $CaCO_3$	$= 40 + 12 + (3 \times 16)$	$= \mathbf{100}$
M_r of CaO	$= 40 + 16$	$= \mathbf{56}$
M_r of CO_2	$= 12 + (2 \times 16)$	$= \mathbf{44}$
Atom economy	$= \frac{56}{100} \times 100$	$= \mathbf{56\%}$

③ Yield

Yield and percentage yield

1 Define the term yield.

....

....

2 What piece of measuring equipment can be used to measure the mass of a desired product?

....

3 a) What is the unit of percentage yield?

....

b) What is the unit of theoretical yield?

....

c) What is the unit of actual yield?

....

4 Write the equation for calculating percentage yield.

....

5 What is the percentage yield of a chemical reaction where 10g of substance was collected and the theoretical yield was 50g?

....

6 What is the actual yield of a chemical reaction where it has a 25% percentage yield and a theoretical yield of 100g?

....

Atom economy

7 What is atom economy also known as?

....

8 Water can be made from its elements:

$2H_2 + O_2 \rightarrow 2H_2O$

What is the atom economy for this reaction?

....

🔴 Volume of gas under standard conditions

It is important to have scientific standards so that **data** can be collected in the same situations and can be used to draw **valid conclusions**.

Standard conditions are:
- temperature = 20°C
- pressure = 1 atmosphere = 100 kPa.

One mole of any gas under standard conditions will take up 24 dm³ (24 litres or 24 000 cm³) of space.

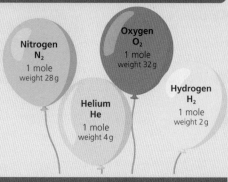

Nitrogen
N_2
1 mole
weight 28 g

Oxygen
O_2
1 mole
weight 32 g

Hydrogen
H_2
1 mole
weight 2 g

Helium
He
1 mole
weight 4 g

🔴 Calculating volume of gas

It is possible to calculate the volume that a gas will occupy under standard conditions using this formula:

> Volume (dm³) = amount of gas (mol) × 24

For any gas under standard conditions:
- 1 mole has a volume of 24 dm³
- 0.5 mole has a volume of 12 dm³
- 2 moles have a volume of 48 dm³.

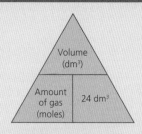

Volume
(dm³)

Amount
of gas
(moles)

24 dm³

🔴 Volume of gas in balanced symbol equations

Consider the reaction between 0.65 g of zinc and excess hydrochloric acid:

The **balanced symbol** equation for the reaction is:

$Zn(s) + 2HCl(aq) \rightarrow ZnCl_2(aq) + H_2(g)$

Zinc is the **limiting reactant** and we can use this to determine the volume of hydrogen produced.

Determine the volume of hydrogen produced in the reaction between 0.65 g of zinc and excess hydrochloric acid.

Calculate the moles of zinc.

= mass ÷ A_r
= 0.65 ÷ 65
= 0.01 moles

The balanced symbol equation shows that the **ratio** of zinc : hydrogen is 1 : 1. So, there will be 0.01 moles of hydrogen produced.

Now, calculate the volume of hydrogen gas produced under standard conditions.

Volume of hydrogen = 0.01 × 24 = 0.24 dm³

Although the volume of any gas is the same under standard conditions, the mass of the gas would be different. For example, under standard conditions:
- 1 mole of helium gas would have a volume of 24 dm³ and a mass of 4 g.
- 1 mole of argon gas would also have a volume of 24 dm³ but a mass of 40 g.

This is because each atom of argon has a greater mass than each atom of helium.

③ Volume of gas

Volume of gas under standard conditions

1 **a)** What is the temperature in standard conditions?

b) What is the pressure in standard conditions?

2 What volume does one mole of gas take up under standard conditions?

3 How do you change dm^3 to cm^3?

Calculating volume of gas

4 Write the formula to calculate the volume of a gas under standard conditions.

5 **a)** How many moles of gas, under standard conditions, occupy $48\,dm^3$?

b) How many moles of gas, under standard conditions, occupy $24\,dm^3$?

c) How many moles of gas, under standard conditions, occupy $12\,dm^3$?

Volume of gas in balanced symbol equations

6 What gas is made when zinc reacts with hydrochloric acid?

7 Give the state symbol for showing a substance is in the gas state.

8 How many moles of hydrogen gas are made when 0.5 moles of zinc is fully reacted with hydrochloric acid?

9 What volume of hydrogen gas is made, under standard conditions, when 0.5 moles of zinc is fully reacted with hydrochloric acid?

③ Chemical and physical changes

Chemical and physical change

Chemical changes:
- are not easily reversible
- involve a new substance being made
- have no change in mass.

Physical changes:
- are reversible
- involve no new substance being made
- have no change in mass.

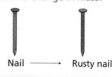

Nail ⟶ Rusty nail

Butter ⟶ Butter melts

Gas formation

Energy transfer (light, heat)

Chemical reaction observations

Colour change

Precipitate formation

Equations

Word equations give the names of the reactants and the products that are formed in a chemical reaction. Word equations should:
- include the names of all the chemicals involved in the reaction
- have an arrow (⟶) to show that the reactants go to the products
- have all the reactants on the left side of the arrow
- have all the products on the right side of the arrow.

> Never use an equals sign (=) in a chemical reaction: the reactants do not equal the products, they change into the products.

For example: **hydrogen + oxygen ⟶ water**

Consider the reaction to make water from its elements:

The reactants are two molecules of hydrogen and one molecule of oxygen.

The bonds in the reactants break to form two atoms of oxygen and four atoms of hydrogen.

The atoms rearrange and make new bonds to form the products which are two molecules of water.

$2H_2 + O_2 \rightarrow 2H_2O$
Two molecules of hydrogen ($2H_2$) react with one molecule of oxygen (O_2) to make two molecules of water ($2H_2O$).

HT Ionic equations show only the particles that are doing the chemistry. Ionic equations only have:
- the formulae of the **dissolved ions** directly involved in the chemical reaction
- the formulae of other **compounds** or **elements** that are not dissolved.

In the **neutralisation** reaction between hydrochloric acid and sodium hydroxide:
- Word equation: hydrochloric acid + sodium hydroxide ⟶ sodium chloride + water
- Balanced symbol equation: $HCl(aq) + NaOH(aq) \rightarrow NaCl(aq) + H_2O(l)$
- Ions in solution: $H^+(aq) + Cl^-(aq) + Na^+(aq) + OH^-(aq) \rightarrow Na^+(aq) + Cl^-(aq) + H_2O(l)$
- Ionic equation (without the spectator ions): $H^+(aq) + OH^-(aq) \rightarrow H_2O(l)$

> Spectator ions are the ions that are unchanged in the solution where a reaction is happening.

Half equations show either **reduction** or **oxidation** reactions. They are balanced in terms of particles and charge.

In a **displacement** reaction between magnesium and copper sulfate:
- Word equation: zinc + copper sulfate ⟶ zinc sulfate + copper
- Balanced symbol equation: $Zn(s) + CuSO_4(aq) \rightarrow ZnSO_4(aq) + Cu(s)$
- Ions in solution: $Zn(s) + Cu^{2+}(aq) + SO_4^{2+}(aq) \rightarrow Zn^{2+}(aq) + SO_4^{2+}(aq) + Cu(s)$
- Ionic equation: (without the spectator ions): $Zn(s) + Cu^{2+}(aq) \rightarrow Zn^{2+}(aq) + Cu(s)$
- Half equation (for oxidation): $Zn(s) \rightarrow Zn^{2+}(aq) + 2e^-$
- Half equation (for reduction): $Cu^{2+}(aq) + 2e^- \rightarrow Cu(s)$

> If you add the half equation for oxidation with the half equation for reduction, you will get the balanced ionic equation for the redox reaction.

Chemical and physical changes

Chemical and physical change

1 Why is combustion (burning) classified as a chemical change?

....

2 Why is melting (changing state) classified as a physical change?

....

Equations

3 **a)** What are the substances at the start of a change called?

....

b) What are the substances at the end of a change called?

....

4 What does → mean in a word equation?

....

5 Why do symbol equations need to be balanced?

....

....

6 Balance the following chemical equation:

.... Mg + ...O_2 → .... MgO

7 **HT** What information do ionic equations give?

....

....

8 **HT** Define the term spectator ion.

....

....

9 **HT** What do half equations show?

....

10 **HT** Write a half equation for the oxidation of copper atoms to form copper 2^+ ions.

....

Types of chemical reaction

Oxidation and reduction

Oxidation is an **exothermic** (gives out energy) chemical reaction. Oxidation is:
- addition of oxygen, e.g.
  magnesium + oxygen → magnesium oxide

HT • loss of **electrons**, e.g. $Mg \rightarrow Mg^{2+} + 2e^-$

Oxidation reactions include **combustion** and rusting.

Reduction is:
- loss of oxygen, e.g.
  iron oxide + carbon → iron + carbon dioxide

HT • gain of electrons, e.g. $2O^{2-} \rightarrow O_2 + 4e^-$

Reduction reactions include the extraction of metals from their ores.

> When one substance is oxidised, it does so by reducing the other. So, the whole reaction can be described as a REDOX reaction, because reduction and oxidation happen at the same time.

Neutralisation

Neutralisation is usually an exothermic chemical reaction. It is the reaction between an **acid** and a **base**, e.g.
- metal + acid → metal salt + hydrogen
- metal oxide + acid → metal salt + water
- metal hydroxide + acid → metal salt + water
- metal carbonate + acid → metal salt + water + carbon dioxide

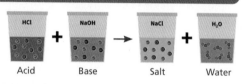

Acid   Base   Salt   Water

Neutralisation always produces a salt, e.g.
- hydrochloric acid makes chlorides
- sulfuric acid makes sulfates
- nitric acid makes nitrates.

Decomposition, precipitation and displacement

Decomposition reactions are **endothermic** (take in energy). There are two types of decomposition reaction:

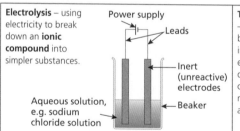

Electrolysis – using electricity to break down an **ionic compound** into simpler substances.

Power supply

Leads

Inert (unreactive) electrodes

Beaker

Aqueous solution, e.g. sodium chloride solution

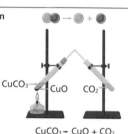

Thermal decomposition – using heat energy to break down a substance into simpler substances, e.g. the thermal decomposition of copper carbonate to make copper oxide and carbon dioxide.

$CuCO_3$ CuO CO_2

$$CuCO_3 \rightarrow CuO + CO_2$$

Precipitation reactions are where two different solutions react to make a solid **insoluble precipitate**. This can be used as an analytical test to show the presence of ions as well as a method for making an insoluble salt. The precipitate can be removed from the reaction mixture by **filtering**.

Displacement reactions happen when a more reactive element takes the place of a less reactive element in its compound. Here are two examples:
- A more reactive halogen will take the place of a less reactive halogen in a compound.
- A more reactive metal will take the place of a less reactive metal in a compound.

> Some chemical reactions can be classified as more than one type of reaction, e.g. metal displacement is also a REDOX reaction as the more reactive metal is being oxidised and the less reactive metal is being reduced.

Types of chemical reaction

Oxidation and reduction

1 Why is combustion (burning) classified as an oxidation reaction?

2 **HT** What happens in terms of electrons when a substance is oxidised?

3 Why is extraction of iron from iron ore (iron oxide) classified as a reduction reaction?

4 **HT** What happens in terms of electrons when a substance is reduced?

Neutralisation

5 What are the **two** types of reactants in a neutralisation reaction?

6 What is always made in a neutralisation reaction?

Decomposition, precipitation and displacement

7 Name the **two** types of decomposition.

8 What sort of substance can undergo electrolysis?

9 Define the term precipitate.

10 How can you separate the solid product in a precipitation reaction?

11 Describe what happens in a halogen displacement reaction.

12 Describe what happens in a metal displacement reaction.

4 The reactivity series

Metal reactions

Most **elements** are **metals**. When metal atoms react in a **chemical reaction** they:
- lose their outer shell **electrons**
- become positive **ions**
- are **oxidised**.

> The reactivity of a metal is related to how easily it can become a positive ion.

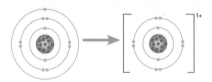

Sodium atom forming a sodium ion by losing an electron

The reactivity series

The **reactivity** of metals is a list of metals with:
- the most reactive metals at the top to the least reactive metals at the bottom
- the metals that form ions most easily at the top to the least at the bottom
- the most easily oxidised metals at the top.

The order of the metals in the reactivity series is based on the chemical reactions of metals with water and acids at room temperature (20°C).

Often the **non-metals** carbon and hydrogen are added to the reactivity series to predict reactions of metals:
- All metals above hydrogen in the reactivity series react with metals and those below do not.
- All metals above carbon in the reactivity series cannot be **extracted** from their **ores** by carbon.

Most reactive

Potassium
Sodium
Calcium
Magnesium
Aluminium
Carbon
Zinc
Iron
Tin
Lead
Hydrogen
Copper
Silver
Gold
Platinum

Least reactive

Displacement reactions

Metal displacement reactions happen when a more reactive metal takes the place of a less reactive metal from its compounds.

> magnesium + iron sulfate → magnesium sulfate + iron
> Mg + Fe SO₄ → Mg SO₄ + Fe

The reactivity series can be used to predict the outcome of metal displacement reactions:
- Underline the names of the metals (pure metal and the metal in a compound) in the reactants.
- The metal that is highest in the reactivity series will be in the compound.
- If the most reactive metal is already in the compound in the **reactants**, there will be no reaction.

When copper is put in a solution of zinc sulfate, there is no reaction. This is because zinc is more reactive than copper and zinc is already in the compound.

> If the most reactive metal is a pure metal in the reactants, there will be a reaction.

When zinc is put in a solution of copper sulfate, there is a displacement reaction. This is because zinc is more reactive than copper and zinc **displaces** (takes the place of) copper from its compound.

> Metal atoms that are large lose their outer shell electrons more easily and are more reactive than small metal atoms. So, the reactivity of metals increases down a group.

④ The reactivity series

Metal reactions

1 What happens to the outer shell electrons when a metal atom reacts?

2 **HT** Write a half equation for the oxidation of a sodium atom.

The reactivity series

3 What is the reactivity series?

4 What chemical reactions are used to create the reactivity series?

5 Which **two** non-metals are often included in the reactivity series?

6 Give an example of a metal that will not react with water but will react with an acid.

Displacement reactions

7 Describe what happens in a metal displacement reaction.

8 What are the products when magnesium reacts with zinc sulfate?

9 Explain why there is no reaction when zinc is put into a solution of magnesium sulfate.

10 Write a word equation for the reaction of copper with magnesium sulfate.

11 Write a word equation for the reaction of magnesium with copper sulfate.

The extraction of metals

Metal as a resource

Metals are a **finite resource**. Metal ores and **unreactive** metals are mined and extracted, which causes **pollution** and habitat damage. But the negative impact of a mine can be reduced:

* Gases can be treated and **neutralised**.
* **Toxic** material can be specially stored to prevent pollution.
* Old mines can be made into nature reserves.

Unreactive metals like gold are found uncombined in nature. But most metals are found in metal **compounds** known as **minerals**, which need to undergo a **reduction** reaction to get the metal.

When it is economical to extract a metal from a mineral, the rock is described as an **ore**.

Reduction with carbon

Metals below carbon in the **reactivity series** can be reduced by carbon. The metal compound is mixed with carbon and heated. The reaction produces the metal and carbon dioxide (linked to **global warming** and **climate change**).

Haematite is the main iron ore containing iron oxide. Iron oxide undergoes carbon reduction in the **blast furnace**, where it reacts to make carbon monoxide, which then reduces the iron oxide to form impure iron. Limestone is also added to react with any acidic impurities to make a substance called **slag**. Slag can be used for breeze blocks and road building.

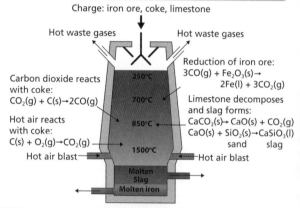

Charge: iron ore, coke, limestone

Hot waste gases

Carbon dioxide reacts with coke:
$CO_2(g) + C(s) \rightarrow 2CO(g)$

Hot air reacts with coke:
$C(s) + O_2(g) \rightarrow CO_2(g)$

Hot air blast

Reduction of iron ore:
$3CO(g) + Fe_2O_3(s) \rightarrow 2Fe(l) + 3CO_2(g)$

Limestone decomposes and slag forms:
$CaCO_3(s) \rightarrow CaO(s) + CO_2(g)$
$CaO(s) + SiO_2(s) \rightarrow CaSiO_3(l)$
sand   slag

Hot air blast

250°C
700°C
850°C
1500°C
Molten Slag
Molten iron

> Reduction with carbon gives an impure form of the metal, so electrolysis can purify the metal.

Electrolysis

Electrolysis is the use of electricity to break down an ionic compound. It is used to extract metals when the metal is above carbon in the reactivity series and would react with carbon. The extraction process needs the metal-containing compound to be melted and an **electric current** passed through. Electrolysis produces a pure metal at the **cathode** and non-metals at the **anode**. Aluminium is extracted using electrolysis:

* Overall balanced symbol equation:
$2Al_2O_3 \rightarrow 4Al + 3O_2$

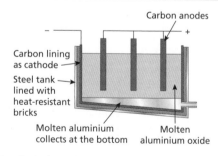

Carbon anodes

Carbon lining as cathode

Steel tank lined with heat-resistant bricks

Molten aluminium collects at the bottom

Molten aluminium oxide

* At the cathode (**oxidation**): $Al^{3+} + 3e^- \rightarrow Al$
* At the anode (**reduction**): $O^{2-} \rightarrow 2O_2 + 4e^-$

The carbon anode needs to be continually replaced as the oxygen gas quickly reacts with the carbon of the anode: $C + O_2 \rightarrow CO_2$

Bauxite is the main ore used to extract aluminium. The ore is processed to remove aluminium oxide, which is then mixed with **cryolite** to form the **electrolyte**. This has a lower melting point than a pure electrolyte so reduces the energy cost.

> Electrolysis is more expensive than metal extraction using carbon, because it needs a lot of energy to melt the ionic compound.

(4) The extraction of metals

Metal as a resource

1 How are unreactive metals found in nature?

2 What is a mineral?

3 What is an ore?

4 Name the chemical reaction used to extract a metal from a metal compound.

Reduction with carbon

5 When can carbon be used to extract a metal from its compound?

6 Write a word equation for the extraction of iron from iron oxide using carbon.

7 Give the environmental problems linked to carbon dioxide.

Electrolysis

8 When is electrolysis used to extract metals?

9 Why is electrolysis an expensive method of metal extraction?

10 Write a word equation for the electrolysis of aluminium oxide.

11 Why does the carbon anode in aluminium electrolysis need to be continuously replaced?

④ REDOX reactions

Reduction

Reduction reactions happen when:
- oxygen is lost by a substance
- electrons are gained by a substance.

Examples of reduction reactions are:
- metal extraction with carbon or electrolysis
- displacement reactions

– in metal displacement reactions, the metal ion in the compound is reduced to form a metal atom
– in halogen displacement reactions, the halogen atom is reduced to form a halide ion.

Oxidation

Oxidation reactions happen when:
- oxygen is gained by a substance
- electrons are lost from a substance.

Examples of oxidation reactions are:
- **combustion**
- **rusting**
- displacement reactions:
  - in metal displacement reactions, the metal atoms are oxidised to form metal ions in the compound
  - in halogen displacement reactions, the halide ions are oxidised to form halogen atoms.
- metals are oxidised to form metal ions in a salt when they react with acids.

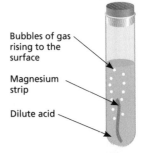

Bubbles of gas rising to the surface

Magnesium strip

Dilute acid

Magnesium atoms are oxidised to magnesium ions in this reaction

REDOX

Reduction and oxidation do not happen on their own: when one substance is oxidised, the other substance is reduced. So, **REDOX** reactions are reactions where oxidation and reduction are happening at the same time.

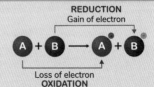

REDUCTION
Gain of electron

A + B ⟶ A + B

Loss of electron
OXIDATION

REDOX reactions include the following:
- Metal displacement reactions, e.g. the REDOX reaction between silver nitrate and copper:
  - The copper displaces the silver from silver nitrate – the **balanced symbol equation** is:
    $Cu(s) + 2AgNO_3(aq) \rightarrow Cu(NO_3)_2(aq) + 2Ag(s)$
  - As copper atoms lose electrons to become copper ions, copper being oxidised can be described using a **half equation**:
    $Cu \rightarrow Cu^{2+} + 2e^-$
  - As silver ions gain electrons to become silver atoms, silver being reduced can be described using a half equation: $Ag^+ + e^- \rightarrow Ag$

- Halogen displacement reactions, e.g. a reaction between potassium bromide and chlorine:
  - The chlorine displaces the bromide from potassium bromide – the balanced symbol equation is:
    $Cl_2(aq) + 2KBr(aq) \rightarrow 2KCl(aq) + Br_2(aq)$
  - As chlorine atoms gain electrons to become chloride ions, chlorine being reduced can be described using a half equation: $Cl_2 + 2e^- \rightarrow 2Cl^-$
  - As bromide ions lose electrons to become bromine atoms, which then form a diatomic molecule, bromide ions being reduced can be described using a half equation: $2Br^- \rightarrow Br_2 + 2e^-$

Oxidants are chemicals that supply oxygen and allow other substances to oxidise more easily. They can be hazardous and have a safety warning symbol. Oxidants undergo a REDOX reaction where they themselves are reduced.

④ REDOX reactions

Reduction

1 What happens, in terms of oxygen, in a reduction reaction?

2 **HT** What happens, in terms of electrons, in a reduction reaction?

3 Why is electrolysis of aluminium oxide classified as a reduction reaction?

4 Which substance is reduced in the reaction between magnesium metal and hydrochloric acid?

Oxidation

5 What happens, in terms of oxygen, in an oxidation reaction?

6 **HT** What happens, in terms of electrons, in an oxidation reaction?

7 Why is combustion of charcoal (impure carbon) classified as an oxidation reaction?

HT REDOX

8 What is a REDOX reaction?

9 Explain why a displacement reaction is classified as a REDOX reaction.

10 **a)** In the reaction between magnesium metal and copper(II) sulfate, which substance is being oxidised?

b) In the reaction between magnesium metal and copper(II) sulfate, which substance is being reduced?

c) Why is the reaction between magnesium metal and copper(II) sulfate classified as a REDOX reaction?

4 Reactions of acids

Acids

Acids are substances that:
- have a pH <7
- release a $H^+(aq)$ **ion** in solution.

Three common acids you should know are:
- hydrochloric acid, $HCl(aq)$
- sulfuric acid, $H_2SO_4(aq)$
- nitric acid, $HNO_3(aq)$.

Metals and acids

Metals above hydrogen in the **reactivity series** can react with acids to make a metal salt and hydrogen. The general equation for this reaction is:

> **metal + acid → metal salt + hydrogen**

The name of the **metal salt** is found by looking at the name of the acid:

Acid	Salt
Hydrochloric acid	Metal chloride
Sulfuric acid	Metal sulfate
Nitric acid	Metal nitrate

In this reaction, metal atoms are **oxidised** to metal ions and **pH** increases. **Metal oxides** are **basic** and usually **insoluble**. A **neutralisation** reaction can take place between a metal oxide and an acid to produce metal salts and water. The general equation for this reaction is:

> **metal oxide + acid → metal salt + water**

Metal hydroxides are basic and often they dissolve in water, so are **alkalis** too. A neutralisation reaction can happen between a metal hydroxide and an acid to make a metal salt and water. The general equation for this reaction is:

> **metal hydroxide + acid → metal salt + water**

> **acid + base → salt + water**

Metal carbonates are basic and usually insoluble. A **neutralisation** reaction can happen between a metal carbonate and an acid to produce a metal salt, water and carbon dioxide. The general equation for this reaction is:

> **metal carbonate + acid → metal salt + water + carbon dioxide**

Salts

Salts are **ionic compounds** made when the hydrogen in an acid is swapped for a metal ion or the ammonium ion. The **solubility** of salts varies:
- Insoluble salts can be made from **precipitation** reactions and the salt can be separated by **filtering**.
- **Soluble** salts can be made from neutralisation reactions between an acid and an insoluble base (reactive metals, metal oxides, hydroxides or carbonates). **Excess** solid base is added to ensure the reaction is complete, then the filtrate is collected and the salt **crystalised**.

When salts are made, they are impure as they contain some solvent from the reaction mixture. To make a pure dry sample, the salt must be patted dry with absorbent paper or left in a drying oven.

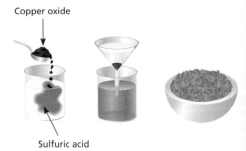

Copper oxide

Sulfuric acid

Add copper oxide to sulfuric acid ⟹ Filter to remove any unreacted copper oxide ⟹ Evaporate to leave behind blue crystals of the 'salt' copper sulfate

④ Reactions of acids

Acids

1 Which ion do acids release in aqueous solution?

....

2 Name the acid with the formula H_2SO_4.

....

Metals and acids

3 Which metals can react with acids?

....

4 Name the chemical reaction between metal oxides and acids.

....

5 What are the names of the products when calcium oxide reacts with sulfuric acid?

....

6 What are the names of the products when sodium hydroxide reacts with hydrochloric acid?

....

7 Write a word equation for the reaction between potassium hydroxide and sulfuric acid.

....

....

8 What gas is made when a metal carbonate reacts with an acid? Tick the correct answer.

Carbon monoxide ☐   Oxygen ☐

Carbon dioxide ☐   Hydrogen ☐

9 Write a word equation for the reaction between sodium carbonate and nitric acid.

....

....

Salts

10 Define the term salt.

....

....

....

11 What chemical reaction can you use to make an insoluble salt?

....

4) The pH scale

The pH scale

The **pH scale** is a measure of acidity or alkalinity of a **solution**. The scale starts at 0 and goes up to 14. The pH can classify a substance:

- **acids** have a pH <7
- **alkalis** have a pH >7
- **neutral solutions** have a pH = 7.

> The pH is a measure of the **concentration** of acid particles or hydrogen **ions** ($H^+(aq)$) in solution. The higher the **concentration** of these ions, the lower the pH value.

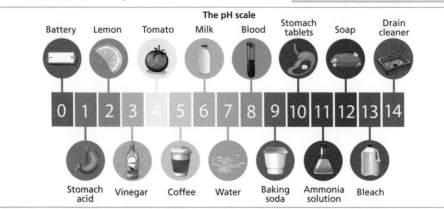

The pH scale

Battery Lemon Tomato Milk Blood Stomach tablets Soap Drain cleaner

0 1 2 3 4 5 6 7 8 9 10 11 12 13 14

Stomach acid Vinegar Coffee Water Baking soda Ammonia solution Bleach

The pH scale is a **logarithmic scale**. This means that a solution with a pH of 1 has:

- × 10 more concentration $H^+(aq)$ when compared to a solution with a pH of 2
- × 100 more concentration $H^+(aq)$ when compared to a solution with a pH of 3
- × 1000 more concentration $H^+(aq)$ when compared to a solution with a pH of 4.

The pH of a solution can be changed by adding water and diluting the solution or by adding a different substance to neutralise some of the acid or alkali.

Measuring pH

pH can be measured with **universal indicator**:

- Add a few drops of universal indicator to the solution or dip the universal indicator paper into the solution.
- Compare the colour of the universal indicator to the colour chart and read off the pH value.

pH can be measured with a **pH probe**:

- Put the pH probe into the solution.
- Read off the pH value from the display.

Neutralisation

A **neutralisation** reaction happens when a base reacts with an acid. Alkalis are soluble **bases**. When an alkali reacts, the acid $H^+(aq)$ reacts with the alkali $OH^-(aq)$ to produce water.

> This can be summarised in an **ionic equation**:
>
> $$H^+(aq) + OH^-(aq) \rightarrow H_2O(l)$$

A pH probe is a measuring instrument. If it is connected to a computer which can store the data, it would be called a datalogger.

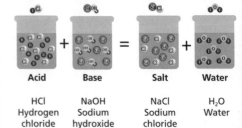

Acid + Base = Salt + Water

HCl
Hydrogen chloride

NaOH
Sodium hydroxide

NaCl
Sodium chloride

H_2O
Water

The pH scale

1 Describe the pH scale.

....

....

2 Give the range that pH can be measured in.

3 How can you change the pH of a solution?

....

....

4 Draw lines to match the key term on the left with the pH value on the right.

Alkali		<7
Acid		>7
Neutral		=7

Measuring pH

5 Which substance can be used to measure the pH of a solution?

....

6 Which measuring instrument can be used to measure the pH of a solution?

....

Neutralisation

7 What sort of substance releases $H^+(aq)$?

....

8 What sort of substance releases $OH^-(aq)$?

....

9 What is made when $H^+(aq)$ and $OH^-(aq)$ react?

....

Titration

A **titration** is an analytical technique that allows you to measure the **concentration** of an **acid** or an **alkali**. You need to know:

- that the acid will react with the alkali in a **neutralisation** reaction
- the **balanced symbol equation** for the reaction
- the **volume** and **concentration** of either the acid or the alkali
- the volume of the unknown acid or alkali.

Acid–base indicators like methyl orange or phenolphthalein are used to show the end point of the titration; they are one colour in acids and a different colour in bases. The colour change is very clear, leading to more accurate and reliable results.

In a titration, the starting volume and the final volume must be measured on the burette. Use this to calculate the volume of solution added from the burette or the **titre**.

It is important to read the burette at eye level and from the bottom of the **meniscus** in order to get an accurate reading.

Results tables and calculations

For a titration, a rough titration is completed to help estimate the volume of solution that should be added from the **burette**. This value is not used in any calculations. It is important that:

- each reading of the burette must be to the same **precision** (to two decimal places)
- **accurate** titrations are completed until there are two **titres** that are **concordant** (within 0.1 cm³ of each other).

The aim of a titration is to calculate the concentration of a solution. Use the **mean** (average) titre for these calculations.

To calculate the mean titre, use the equation:

Mean titre = sum of the two concordant results ÷ 2

	Rough	1st titration	2nd titration	3rd titration
Final burette reading / cm³	37.60	36.20	39.15	38.40
Initial burette reading / cm³	1.80	0.00	3.95	2.10
Titre / cm³	35.80	36.20	35.20	36.30

▦ Example

Break down the calculation:

1. Write down a balanced equation for the reaction to determine the ratio of moles of acid to alkali involved.
2. Calculate the number of moles in the solution of known volume and concentration. You can work out the number of moles in the other solution from the balanced equation.
3. Calculate the concentration of the other solution.

A titration is carried out and 0.04 dm³ hydrochloric acid neutralises 0.08 dm³ sodium hydroxide of concentration 1.00 mol / dm³.

Calculate the concentration of the hydrochloric acid.

$$HCl + NaOH \rightarrow NaCl + H_2O$$ ◄──── Write the balanced symbol equation for the reaction.

Number of moles of NaOH = volume × concentration

$$= 0.08 \, dm^3 \times 1.00 \, mol / dm^3 = 0.08 \, mol$$

Concentration of HCl = $\dfrac{\text{number of moles of HCl}}{\text{volume of HCl}}$

$$= \frac{0.08 \, mol}{0.04 \, dm^3} = 2.00 \, mol / dm^3$$

The balanced equation shows that the amount of hydrochloric acid is equal to the amount of sodium hydroxide, i.e. 0.08 mol.

(4) Titration

Titration

1 What is the aim of a titration?

2 What do you measure in a titration?

3 What is a titre?

4 Describe how to read a burette.

Results tables and calculations

5 **a)** What does accurate mean?

b) What does precise mean?

c) What does concordant mean?

6 Why is a rough titration completed?

7 What is a mean titre?

8 How do you calculate a mean titre?

9 HT Which equation do you use to calculate the number of moles of solute in a solution?

10 HT What information can a balanced symbol equation give for titration calculations?

4 Strong and weak acids

Ionisation

Ionisation is the process of making an **ion**.

$$H\overset{\frown}{-}A \longrightarrow H^\oplus + A^\ominus$$

When **acids** are put into water, they release $H^+(aq)$ ions in **solution** and this is an example of ionisation. The more hydrogen ions there are in a solution, the lower the **pH** value.

This can be achieved by:
- having a more **concentrated** solution
- choosing an acid which is easier to ionise – the strength of the bond between the H and the rest of the substance varies with each acid (the easier it is to break the bond, the greater the ionisation).

Strong acids

Strong acids fully ionise in solution. This means that every acid formula dissociates into a $H^+(aq)$ ion and a negative counter ion. The **general equation** for this ionisation is: **$HA \rightarrow H^+ + A^-$** (where HA is the formula of the acid, H^+ is the acid particle and A^- is the counter ion).

There are three strong acids that you need to know and show the ionisation for:
- Hydrochloric acid, $HCl \rightarrow H^+ + Cl^-$
- Nitric acid, $HNO_3 \rightarrow H^+ + NO^{3-}$
- Sulfuric acid, $H_2SO_4 \rightarrow 2H^+ + SO_4^{2-}$

Strong and concentrated

Strong and dilute

Weak acids

Weak acids like citric acid only partially ionise in aqueous solution. The general equation for the ionisation of a weak acid is: **$HA \rightleftharpoons H^+ + A^-$**

There are two weak acids that you need to know and show the ionisation for:
- Ethanoic acid, $CH_3COOH \rightarrow H^+ + CH_3COO^-$
- Carbonic acid, $H_2CO_3 \rightarrow H^+ + HCO_3^-$

Weak and concentrated

Weak and dilute

Concentration is a measure of how much solute is in a given volume of solution, whereas strength is a measure of how much a substance ionises. So, it is possible to have a concentrated solution of a weak acid or a dilute solution of a strong acid.

Ethanoic acid

Carbonic acid
H_2CO_3

(H) Hydrogen

(C) Carbon

(O) Oxygen

4 Strong and weak acids

Ionisation

1 What is ionisation?

2 How does the amount of $H^+(aq)$ affect pH?

Strong acids

3 **a)** What is a strong acid?

b) Give an example of a strong acid.

4 Write the general equation for the ionisation of a strong acid.

5 Write the balanced equation for the ionisation of hydrochloric acid.

Weak acids

6 **a)** What is a weak acid?

b) Give an example of a weak acid.

7 Write the general equation for the ionisation of a weak acid.

8 Write the balanced equation for the ionisation of ethanonic acid.

9 Write the balanced equation for the ionisation of methanoic acid (HCOOH).

Process of electrolysis

Electrolysis is:
- a **chemical change**, as something new is made
- an **endothermic change**, as energy is taken in by the system.

The process uses electricity to break down an **ionic substance** into simpler substances. The electrical **current** causes:
- positive metal **ions** or hydrogen ions to be attracted to the **cathode** (negative electrode)
- negative non-metal ions or hydroxide ions to be attracted to the **anode** (positive electrode).

Ions then become neutral atoms by gaining electrons (**reducing**) at the cathode and losing electrons (**oxidising**) at the anode.

For electrolysis to happen you need:
- **molten** or **aqueous solution** of an ionic compound
- a supply of **direct current** (e.g. from a battery)
- two **electrodes** (usually made of carbon).

You can monitor the electrolysis by having a **lamp** or **ammeter** in the circuit. When the circuit is complete and the current is flowing, the lamp will shine and the ammeter will have a reading.

Positive anode
Negative cathode
Electrolyte

Only when ions are free to move in electrolytes can electrolysis happen. This is because the ions need to carry the charge in the solution and complete the electrical circuit.

Uses of electrolysis

Electrolysis is used for:

Extracting metals from their compounds: metals like aluminium, that are higher than carbon in the **reactivity series**, will be extracted from their compounds using electrolysis.

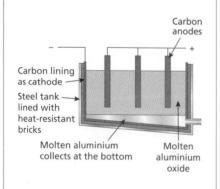

Carbon anodes

Carbon lining as cathode

Steel tank lined with heat-resistant bricks

Molten aluminium collects at the bottom

Molten aluminium oxide

Purifying metals: copper used in electrical wires must be very pure to reduce **resistance**. Impure copper is used to make the anode. The copper atoms become copper ions and enter the copper salt solution. Meanwhile, the copper ions in solution gain electrons at the cathode and become pure copper metal. The cathode is used to make the electrical wires.

Electrolysis of copper(II) sulfate copper purification

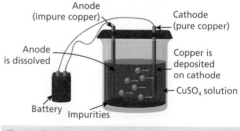

Anode (impure copper)
Cathode (pure copper)

Anode is dissolved

Copper is deposited on cathode

Battery
Impurities
CuSO$_4$ solution

Anode: $Cu \longrightarrow Cu^{2+} + 2e^-$
Cathode: $Cu^{2+} + 2e^- \longrightarrow Cu$

④ Electrolysis

Process of electrolysis

1 a) Why is electrolysis an example of a chemical change?

....

b) Why is electrolysis an example of an endothermic change?

....

2 What sort of ions are attracted to the cathode?

....

3 HT What happens at the anode?

....

....

4 What sort of current is needed for electrolysis?

....

5 What state must the electrolyte be in?

....

6 Which element are electrodes usually made from?

....

Uses of electrolysis

7 Name the metals that are extracted from their compounds using electrolysis.

....

....

8 In the extraction of aluminium, which electrode is the aluminium formed at?

....

9 Complete the sentence by filling in the gaps.

In the purification of copper, the mass of the anode .... and the mass of the

cathode ....

4 Predicting the products of electrolysis

What happens at the anode and the cathode?

The **anode** is positive and attracts the negative **ions** in the **electrolyte**. Ions are **reduced** at the anode.

- In electrolysis of a melt, a **non-metal** is formed. The non-metal atoms will bond together to make **diatomic covalent molecules**.
- In electrolysis of a **solution**:
  - If the ionic compound doesn't contain a **halide**, oxygen is formed,

    HT e.g. $4OH^- \rightarrow O_2 + 2H_2O + 4e^-$

  - If the ionic compound does contain a halide, a **halogen** is formed,

    HT e.g. $2Cl^- \rightarrow Cl_2 + 2e^-$

The **cathode** is negative and attracts the positive ions in the electrolyte. Ions are **oxidised** at the anode.

- In electrolysis of a melt, a **pure metal** is formed.
- In electrolysis of a solution:
  - If the ionic compound contains a metal below hydrogen in the **reactivity series**, then the pure metal is formed,

    HT e.g. $Cu^{2+} + 2e^- \rightarrow Cu$

  - If the ionic compound contains a metal above hydrogen in the reactivity series then hydrogen is formed,

    HT e.g. $2H^+ + 2e^- \rightarrow H_2$

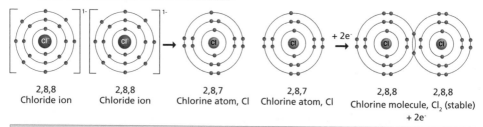

| 2,8,8 Chloride ion | 2,8,8 Chloride ion | 2,8,7 Chlorine atom, Cl | 2,8,7 Chlorine atom, Cl | 2,8,8 Chlorine molecule, Cl_2 (stable) 2,8,8 + 2e⁻ |

Each electrode attracts the oppositely charged ions from the electrolyte. Only one ion at each electrode can be discharged as an element. The reactivity of the elements determines what is produced.

Electrolysis of lead bromide

Molten lead bromide can be electrolysed in conditions available in a school lab as the melting point of lead bromide can be achieved using a Bunsen burner.

The products of this reaction are:

- at the anode, bromide ions are oxidised to bromine atoms:

  HT $2Br^- \rightarrow Br_2 + 2e^-$

- at the cathode, lead ions are reduced to molten lead:

  HT $Pb^{2+} + 2e^- \rightarrow Pb$

Other products

In electrolysis of melts, the temperatures are so high that carbon electrodes can react with the non-metals that are formed, e.g. in aluminium extraction, the carbon anode burns away as it reacts with the oxygen formed to make carbon dioxide.

In electrolysis of solutions, the **spectator ions** form a third product. So, in the electrolysis of brine (NaCl(aq)), chlorine gas is formed at the anode, hydrogen gas at the cathode and a solution of sodium hydroxide is formed as well.

Predicting the products of electrolysis

What happens at the anode and the cathode?

1 Complete the sentences by filling in the gaps.

Non-metal substances are always formed at the ....

If a solution of a metal halide is electrolysed, a .... is formed at the anode.

If a solution of a metal sulfate is electrolysed, ... is formed at the anode.

2 **HT** Write a half equation for the production of bromine at the anode during the electrolysis of potassium bromide.

....

3 What type of substance is always formed at the cathode in the electrolysis of a melt?

....

4 What is formed at the cathode if a solution of sodium chloride is electrolysed?

....

5 What is formed at the cathode if a solution of a copper sulfate is electrolysed?

....

6 **HT** Write a half equation for the production of copper at the cathode during the electrolysis of copper bromide.

....

Electrolysis of lead bromide

7 **a)** Give the formula of the element formed at the cathode during the electrolysis of molten lead bromide.

....

b) Give the formula of the element formed at the anode during the electrolysis of molten lead bromide.

....

Other products

8 Explain why the carbon anodes need to be replaced frequently in the extraction of aluminium with electrolysis.

....

....

....

9 Define the term spectator ion.

....

....

Exothermic reactions

Energy

Energy can be transferred between the **chemical system** and the **surroundings**. The Law of Conservation of energy means:
- energy can neither be created nor destroyed

- the amount of energy in the Universe must remain constant
- the total energy in the system and the surroundings must be the same at the start and end of the change.

Exothermic reactions

Exothermic reactions release energy from the chemical system into the surroundings. So, the total **stored chemical energy** in the **reactants** is more than the total stored energy in the **products**.

The energy transferred to the surroundings causes a rise in **temperature**, which can be observed by using a **thermometer**.

The reaction mixture is put in an insulated vessel to reduce energy being lost to the surroundings. The thermometer measures the temperature change. The bigger the rise:
- the more energy has been released from the reactants as they become the products
- the more exothermic the reaction.

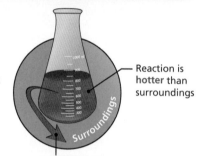

Reaction is hotter than surroundings

Surroundings

Reaction releases heat

Examples of exothermic reactions

Reaction profiles are diagrams that model how the energy changes from the start, during and at the end of the reaction.

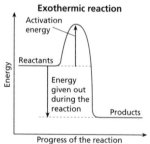

Exothermic reaction

Activation energy

Reactants

Energy

Energy given out during the reaction

Products

Progress of the reaction

In an exothermic reaction:
- reactants have more energy than the products
- the **activation energy** is the hill shape
- the difference between the reactants and the products is the energy released to the surroundings.

Most chemical reactions are exothermic reactions. They include:
- many **oxidation** reactions, including combustion
- **neutralisation**.

Everyday uses of exothermic reactions include:
- Hand warmers, where an exothermic change causes heat lost to the surroundings that can be used to warm a person.
- Self-heating cans that heat up food or drink: a button is pushed, which pierces the foil separator and allows liquid water to mix with solid quicklime. The exothermic reaction happens and a solution of calcium hydroxide is made. The chemical equation for the reaction is: $CaO(s) + H_2O(l) \rightarrow Ca(OH)_2(aq)$.

The heat given out is used to heat the food / drink in the container. (The chemicals from the reaction do not mix with the food / drink.)

The **activation energy** is the minimum energy required to start the reaction. This is usually a spark, flame or heat from friction to get an exothermic reaction to start. The activation energy of a reaction can be reduced by adding a catalyst, which gives an alternative pathway with a lower activation energy, so more collisions are successful and the rate of reaction increases.

Exothermic reactions

Energy

1 How does the total energy in the system and the surroundings compare at the start and end of a change?

Exothermic reactions

2 For an exothermic reaction, what happens to the energy from the chemical system?

3 In an exothermic reaction, how does the stored chemical energy of the reactants compare to the products?

4 What measuring instrument do you use to monitor temperature?

5 **a)** What is a reaction profile?

b) How do you use a reaction profile to calculate the energy released from an exothermic reaction?

c) How do the reactants compare to the products in an exothermic reaction's energy profile?

Examples of exothermic reactions

6 Give **two** everyday uses of exothermic reactions.

7 What is the formula of the liquid reactant used in the self-heating can?

8 What is the state of the product in the self-heating can? Tick the correct option.

Liquid ☐   Gas ☐   Aqueous solution ☐

5 Endothermic reactions

Endothermic reactions

Enothermic reactions do not release energy from the **chemical system**; they absorb energy from the surroundings. So, the total stored chemical energy in the **products** is more than the total stored energy in the **reactants**.

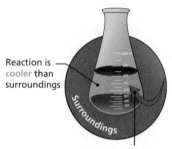

Reaction is cooler than surroundings

Surroundings

Reaction absorbs heat

The energy transferred from the surroundings causes a decrease in **temperature**, which can be observed by using a **thermometer**.

Endothermic reactions can be modelled in a reaction profile.

Endothermic reaction

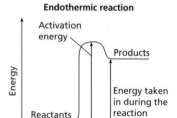

Activation energy

Products

Energy

Energy taken in during the reaction

Reactants

Progress of the reaction

In an endothermic reaction:
• products have more energy than the reactants
• the **activation energy** is from the reactant line to the top of the hill shape
• the difference between the reactants and the products is the energy taken in from the surroundings.

Examples of endothermic reactions

Endothermic reactions include **thermal decomposition** reactions where heat is used to break down a substance. For example, copper(II) carbonate decomposes to copper(II) oxide and carbon dioxide in an endothermic reaction.

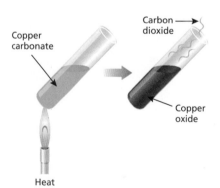

Copper carbonate

Carbon dioxide

Copper oxide

Heat

Everyday uses of endothermic reactions include:
• Sherbet is a formulation made from citric acid and sodium hydrogencarbonate, used in sweets. It mixes with water in the mouth and causes an endothermic neutralisation reaction forming a salt, water and carbon dioxide gas, which makes your mouth feel cool and tingly.
• When athletes injure themselves during sports they can use a cool pack, which contains two chemicals. When the outer bag is crushed, the inner bag of water bursts and mixes with the sodium ammonium nitrate. There is a physical change as the ammonium nitrate dissolves into the water to make a solution of ammonium nitrate. The equation for the change is:

$$NH_4NO_3(s) \xrightarrow{H_2O} NH_4NO_3(aq)$$

Energy is needed from the surroundings for **dissolving** to happen so this is an endothermic reaction.

Endothermic reactions

Endothermic reactions

1 For an endothermic reaction, what happens to the energy from the surroundings?

....

....

2 In an endothermic reaction, how does the stored chemical energy of the reactants compare to the products?

....

....

....

3 What happens to the temperature of the surroundings near an endothermic reaction?

....

4 **a)** What is the x-axis on a reaction profile?

....

b) What is the y-axis on a reaction profile?

....

5 How do you use a reaction profile to calculate the energy released from an endothermic reaction?

....

....

Examples of endothermic reactions

6 Which chemical reaction is endothermic?
Tick the correct answer.

Neutralisation ☐

Thermal decomposition ☐

Oxidation ☐

7 Give **one** everyday use of endothermic reactions.

....

5 Bond energies

Modelling chemical reactions

A **scientific model** can be used to represent the changes happening in a **chemical reaction**:

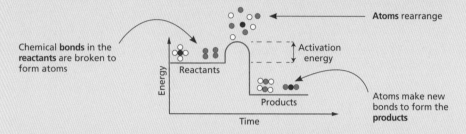

Chemical **bonds** in the **reactants** are broken to form atoms

Atoms rearrange

Activation energy

Atoms make new bonds to form the **products**

Making and breaking bonds

Atoms are held together by chemical bonds.

Breaking bonds is an **endothermic** change. The stronger the bonds, the more energy is needed to break them.

Forming chemical bonds is an **exothermic** change. The stronger the bonds, the more energy is released when they are formed.

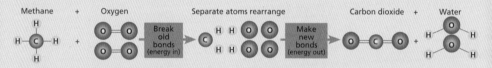

Methane + Oxygen → Separate atoms rearrange → Carbon dioxide + Water

Break old bonds (energy in)

Make new bonds (energy out)

Calculating the energy change of a reaction

The energy change of a reaction can be calculated from the **bond energy** data. Consider the **complete combustion** of methane:
$$CH_4 + 2O_2 \rightarrow CO_2 + 2H_2O$$
Energy needed to break the bonds in the reactants:

Type of bond	Number	Energy (kJ/mol)
C–H	4	4 × 412 = 1648
O=O	2	2 × 496 = 992
	Total	2640

Energy released when the bonds are formed in the products:

Type of bond	Number	Energy (kJ/mol)
C=O	2	2 × 805 = 1610
H–O	4	4 × 463 = 1852
	Total	3462

Overall energy change for a reaction = energy needed to break the reactant bonds – energy released when product bonds are made

2640 – 3462 = -822 kJ/mol

If the overall energy change is a negative number, then energy is released to the surroundings and the reaction is exothermic.

If the overall energy change is a positive number then the energy is taken in from the surroundings and the reaction is endothermic.

5 Bond energies

Modelling chemical reactions

1 **a)** In a chemical reaction, what happens to the bonds in the reactants?

....

b) In a chemical reaction, what happens to the bonds in the products?

....

Making and breaking bonds

2 Complete the sentences by filling in the gaps.

Breaking a bond is an .... change because energy is needed. Making a bond is

an .... change because energy is ....

3 How does the strength of the C=O bond compare to the strength of the C–H bond?

....

Calculating the energy change of a reaction

4 Give the unit that bond energy is measured in.

....

5 **a)** For an exothermic reaction, explain how the strength of all the reactant bonds compare to the strength of all the product bonds.

....

....

b) For an endothermic reaction, explain how the strength of all the reactant bonds compare to the strength of all the product bonds.

....

....

6 Write the formula to calculate the overall energy change of a reaction from bond energy data.

....

....

....

7 **a)** What does a negative energy change for a reaction mean?

....

b) What does a positive energy change for a reaction mean?

....

⑤ Chemical cells

Simple cells and batteries

Chemical cells are a store of chemical energy. Chemicals in the cell react together to produce electricity. A **voltmeter** can be used to measure the voltage (potential difference) produced.

A **simple cell** is made of two different metals, both in contact with an **electrolyte**.

The **voltage** of a cell can be altered by changing the:
- electrode – the biggest voltage is produced when the difference in the **reactivity** of the two metals is the largest
- electrolyte - different ions in the electrolyte can react differently with the metals in the electrodes.

A **battery** is more than one chemical cell connected in **series**. Batteries provide a higher voltage than a single cell.

> The circuit symbol for a cell is: ⊣⊢
>
> As more than one cell joined in series is a battery, the circuit symbol for a battery made of three cells is:
> ⊣|⊣|⊢

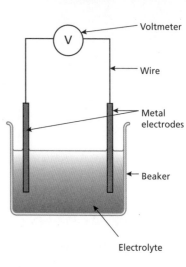

A simple cell

Voltmeter

Wire

Metal electrodes

Beaker

Electrolyte

Non-rechargeable cells

Non-rechargeable cells and batteries:
- are single use
- have limited **reactants**
- have irreversible chemical reactions.

An example is alkaline batteries. Alkaline batteries are cheap to make, but they are expensive to recycle and can cause pollution if disposed of in landfill.

Rechargeable cells

Rechargeable cells and batteries:
- can be used many times
- have chemical reactions in them that can be **reversed** when an external electrical **current** is supplied.

Rechargeable cells can be used many times and so are more **sustainable** as they protect **finite** (non-renewable) resources. However, they cost more to manufacture and are still difficult to recycle so they could cause pollution if disposed of in landfill.

Chemical cells

Simple cells and batteries

1 What is the function (job) of a chemical cell?

....

....

2 Complete the sentence by filling in the gaps.

A simple chemical cell is made of two .... made from two different

...., and an ....

3 What is a battery?

....

Non-rechargeable cells

4 Give an example of a non-rechargeable cell.

....

5 What type of chemical reaction is used in non-rechargeable cells?

....

6 Describe the environmental issues of using non-rechargeable cells.

....

....

7 **a)** What is an economic benefit of using non-rechargeable cells?

....

....

b) Suggest one economic drawback of using non-rechargeable cells.

....

Rechargeable cells

8 What type of chemical reaction is used in rechargeable cells?

....

9 Explain why a rechargeable cell is considered more sustainable than a non-rechargeable cell.

....

....

10 How do you recharge a rechargeable cell?

....

5 Fuel cells

Fuel cells

Fuel cells are an alternative to **chemical cells** and **batteries**. For example, they are being used to replace car engines.

Fuel cells **oxidise** a fuel using **electrochemical reactions**, releasing electricity to produce a continuous **potential difference** (voltage). They need a constant supply of fuel and oxygen.

Hydrogen fuel cells

Hydrogen fuel cells offer an alternative to rechargeable cells and batteries. In hydrogen fuel cells, hydrogen is the fuel that is oxidised to form water.

The overall chemical reaction is:

hydrogen + oxygen → water
$$2H_2 + O_2 \rightarrow 2H_2O$$

HT We can write half **equations** for the reactions at each **electrode**:
- At the negative electrode, **oxidation** happens: $2H_2 + 4OH^- \rightarrow 4H_2O + 4e^-$
- At the positive electrode, **reduction** happens: $O_2 + 2H_2O + 4e^- \rightarrow 4OH^-$

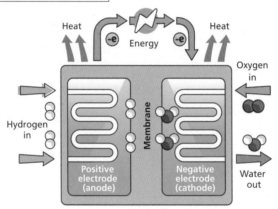

Heat   Energy   Heat
-e   -e
Oxygen in
Membrane
Hydrogen in
Positive electrode (anode)   Negative electrode (cathode)   Water out

Fuel cell

In the cell, an electric **current** happens due to the electrons flowing through an external circuit from the positive electrode (anode) to the negative electrode (cathode).

Hydrogen fuel cells are currently used in space craft. One advantage of this is that the **pure** water which is made by the fuel cell can be used as **potable** water by the astronauts.

Advantages of hydrogen fuel cells	Disadvantages of hydrogen fuel cells
• Quiet and efficient • Easy to maintain as they have no moving parts • Small in size • Make no pollution from the products • **Sustainable**, as hydrogen can be made from the **electrolysis** of water using solar power	• Expensive to manufacture • Hydrogen fuel is a **flammable** gas, which is difficult to store and can be dangerous to store and use

Electrochemical oxidation happens at lower temperatures than combustion oxidation. The chemical energy is converted into electricity for electrochemical oxidation instead of thermal energy (heat) for combustion.

Fuel cells

1 What is the function (job) of a fuel cell?

....

2 What happens to the fuel in a fuel cell?

....

Hydrogen fuel cells

3 What is the product of a hydrogen fuel cell? Tick the correct option.

Hydrogen ☐   Water ☐

Oxygen ☐   Acid ☐

4 Write a balanced symbol equation for the overall chemical reaction in a hydrogen fuel cell.

....

5 The diagram shows a hydrogen fuel cell.

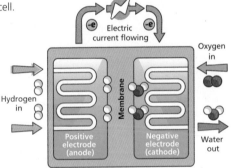

Fuel cell

a) Label the place where oxidation happens in the cell.

b) Label the place where reduction happens in the cell.

6 HT a) Write a half equation for the oxidation reaction in the hydrogen fuel cell.

....

b) Write a half equation for the reduction reaction in a hydrogen fuel cell.

....

7 Why are hydrogen fuel cells considered to be a sustainable source of electricity?

....

....

8 Give **one** economic drawback of hydrogen fuel cells.

....

Measuring rate of reaction

The **rate** of a **chemical reaction** is:
- the speed of the chemical change
- defined by the change in the mass, concentration or volume of the **reactant** used or the mass, concentration or volume of the **product** formed in a given time.

> The mean (average) rate of reaction
> $$= \frac{\text{quantity of reactant used}}{\text{time taken}}$$
> $$= \frac{\text{quantity of product formed}}{\text{time taken}}$$

Some chemical reactions appear to have a **mass** change and can be monitored by using a **top pan balance**. The mass appears to change because:
- a reactant is a gas found in the air, causing the mass to appear to increase
- a product is a gas lost to the air, causing the mass to appear to decrease.

So, rate of reaction would be measured in g/s.

Some chemical reactions produce a gas, which can be collected either by **displacement** of water or in a **gas syringe**. The **volume** is measured, so rate of reaction would be measured in cm³/s. It is also possible to monitor the change in amount of one of the substances. In this case, the unit for rate of reaction would be mol/s.

Some chemical reactions make an **insoluble precipitate**, which makes the reaction mixture **turbid** (cloudy). You can observe a chemical reaction and time how long it takes until you cannot see through the solution. This method is known as the disappearing cross.

Displacement of water

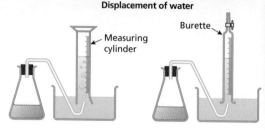

Burette

Measuring cylinder

Using a gas syringe

Gas syringe

The disappearing cross

Timer

Add dilute hydrochloric acid

Flask

Sodium thiosulfate

Paper with cross drawn on it

Calculating rate of reaction

Numerical data can be collected from experiments and plotted on a graph:
- The **independent variable** is time and is plotted on the x-axis.
- The **dependent variable** is either volume or mass and is plotted on the y-axis.
- The **gradient** gives the rate of reaction at that point. The higher the gradient / steeper the curve, the faster the rate of reaction.

The graph shows that reaction A is faster than reaction B.

HT To calculate the gradient of a point on a curve, you first need to draw a tangent to the curve. Then calculate the gradient of the tangent.

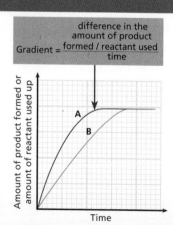

$$\text{Gradient} = \frac{\text{difference in the amount of product formed / reactant used}}{\text{time}}$$

Amount of product formed or amount of reactant used up

A

B

Time

Measuring and calculating rate of reaction

Measuring rate of reaction

1 Explain what is meant by rate of reaction.

....

....

....

2 What measuring instrument do you use to measure mass?

....

3 **a)** Why might the mass of a chemical reaction appear to increase?

....

b) Why might the mass of a chemical reaction appear to decrease?

....

4 What does turbidity mean?

....

5 How could you monitor the rate of reaction between sodium thiosulfate and hydrochloric acid, which produces a cloudy liquid?

....

6 What are the **two** units that rate of reaction can be measured in?

....

Calculating rate of reaction

7 When drawing a graph of rates of reaction data, which axis is always the independent variable?

....

8 What is the relationship between the gradient of a rate of reaction graph and the rate of reaction?

....

....

9 What does it mean when the rate of reaction trend line is horizontal (gradient = 0)?

....

6 Collision theory

Collision theory

Collison theory is a **scientific model** that can be used to explain:
- how chemical reactions happen
- the effect of changing conditions on a chemical reaction.

For a chemical reaction to happen, a collision must happen between the **reactants**, in the correct orientation and with enough energy to be equal to, or higher than, the **activation energy**. So, most collisions do not lead to a reaction.

Increasing rate of reaction

Increasing **temperature** increases the rate of reaction. This is because the particles:
- move faster, increasing the number of collisions per unit time
- have more energy so each collision is more likely to be equal to or higher than the activation energy.

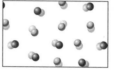 

Cold reaction   Hot reaction

Increasing **surface area** of a solid reactant increases the rate of reaction. This is because:
- more of the solid particles are available for collision at any one time
- the number of collisions in a given time increases, but doesn't affect the percentage of successful collisions.

Large particle   Smaller particles
(small surface area) (larger surface area)

Increasing **concentration** of a reactant in solution increases the rate of reaction. This is because:
- more reactant particles are available for collision at any one time
- the number of collisions in a given time increases, but doesn't affect the percentage of successful collisions
- as a reaction happens, the concentration of the reactants reduces and the concentration of products increases.

Low   High
concentration concentration

🔵 Reacting particle of substance **A**
🔵 Reacting particle of substance **B**

Adding a **catalyst** increases the rate of reaction as:
- it provides an alternative pathway for the reaction to happen
- it reduces the activation energy
- although there is the same number of collisions in a given time, more are successful.

Energy

Reactants

Activation energy without catalyst
Activation energy with catalyst

Products

Progress of reaction

Increasing **pressure** of a gaseous reactant increases the rate of reaction. This is because:
- pressure is like the concentration of a gas, so if the pressure of a gaseous reactant increases, there are the same number of particles available but they are in a smaller volume so more likely to collide.

Increase in pressure increases ➡ rate of reaction

Any reaction has the fastest rate of reaction at the start, slows down and stops at the end.

Different reactions need different catalysts. A catalyst in a **biological system** is an **enzyme**.

Collision theory

Collision theory

1 What is collision theory?

....

....

2 What does collision theory state is needed for a chemical reaction to happen?

....

....

....

Increasing rate of reaction

3 Complete the sentences by filling in the gaps.

When the temperature is increased, the rate of reaction .... and the

overall number of collisions in a given time ....

4 Explain why there are more successful collisions in a given time if surface area of a solid reactant is increased.

....

....

....

....

....

5 What happens to the rate of reaction if you increase the concentration of a reactant in solution?

....

6 What is the name given to a biological catalyst?

....

7 If you add a catalyst to a reaction, what happens to the overall number of collisions in a given time?

....

8 Which diagram shows increased temperature during a reaction?

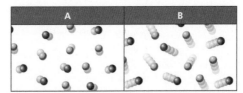

Reversible reactions

Some chemical reactions are described as reversible where the:
- reactants make the products
- products make the reactants.

Reversible reactions are summarised in chemical equations, but instead of having a normal arrow, this arrow is used: $\rightleftharpoons$.

Forward reaction

Backward reaction

The direction of the reversible reaction can be changed by changing the conditions. Conditions include **temperature, amount of substance** and **pressure**.

The energy change of a forward or backward reversible reaction is the same, but either taken in from the surroundings (endothermic) or released to the surroundings (exothermic). For example, hydrated copper(II) sulfate can thermally decompose and takes in 77 kJ/mol of energy; then when anhydrous copper(II) sulfate becomes hydrated, it releases 77 kJ/mol of energy.

Ammonium chloride is a white powder that can thermally decompose to form ammonia and hydrogen chloride gas. As the gases cool, they easily undergo a neutralisation reaction to reform the ammonium chloride salt. So, this is an example of a reversible reaction.

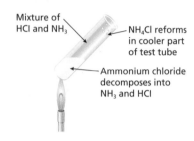

Mixture of HCl and NH_3

NH_4Cl reforms in cooler part of test tube

Ammonium chloride decomposes into NH_3 and HCl

The chemical equations for this reaction are:

$$\text{ammonium chloride} \underset{\text{cool}}{\overset{\text{heat}}{\rightleftharpoons}} \text{ammonia} + \text{hydrogen chloride}$$

$$NH_4Cl(s) \underset{\text{cool}}{\overset{\text{heat}}{\rightleftharpoons}} NH_3(g) + HCl(g)$$

Equilibrium

If reversible reactions happened in a closed system, like a conical flask with a bung in, then:
- no substances can enter
- no substances can leave
- energy can move into and out of the system
- the system will reach equilibrium.

> Catalysts increase the rate of the forward and backward reactions by the same amount. This means that catalysts allow reversible reactions in a closed system to get to equilibrium quicker.

At equilibrium:
- the concentration of all the substances is constant
- rate of forward reaction = rate of reverse reaction.

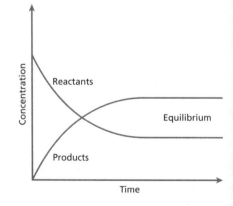

(6) Reversible reactions

Reversible reactions

1 Define reversible reaction.

....

....

....

2 **a)** What is the condition which causes thermal decomposition of ammonium chloride?

....

b) What is the condition which causes the neutralisation reaction between ammonia and hydrogen chloride gases?

....

c) The thermal decomposition of ammonium chloride requires 176 kJ/mol. How much energy is released in the neutralisation reaction between ammonia and hydrogen chloride gas?

....

3 Draw lines to join the boxes.

| The forward reaction of a reversible reaction is exothermic. | The backward reaction is exothermic. |

| The forward reaction of a reversible reaction is endothermic. | The backward reaction is endothermic. |

4 What do you observe when hydrated copper(II) sulfate is heated?

....

Equilibrium

5 **a)** What is needed for a chemical reaction to be at equilibrium?

....

b) How do the rates of the forward and reverse reactions compare when a chemical reaction is at equilibrium?

....

c) How do the concentrations of the substances compare when a chemical reaction is at equilibrium?

....

6 What effect does a catalyst have on a reversible reaction in a closed system?

....

....

The effect of changing conditions on reversible reactions

Le Chatelier's Principle

Conditions for an **equilibrium** reaction can be changed and this causes a change in the relative amounts of the substances. The effect of changing the conditions can be predicted using **Le Chatelier's Principle**. It states that for a **reversible reaction** at equilibrium, the system will oppose any change to the conditions.

> Catalysts have no effect on the position of equilibrium.

Changing conditions

Concentration	Temperature	Pressure
If the concentration of one substance is increased, the reaction will move the position of equilibrium to ensure that the concentration returns back to the original level.	If the temperature is changed, the equilibrium system will favour the reaction to return the temperature back to the original level.	Changes in **pressure** only affect an equilibrium where the substances are in the gas phase.

If you increase the concentration of a **reactant**, the system will oppose the change by increasing the rate of the forward reaction. This lowers the concentration of the reactants and increases the concentration of the **products**.

If temperature is increased:
- the system favours the **endothermic** reaction and increases the rate of this reaction
- the relative amount of product increases for a reaction that is endothermic in the forward direction
- the relative amount of product decreases for a reaction that is **exothermic** in the forward direction

- If pressure increases, the system will increase the rate of reaction for the reaction that has the least amount of gas.

Increase in reactant

Equilibrium shifts to **right**
More A is converted to B

If temperature is decreased:
- the system favours the exothermic reaction and increases the rate of this reaction
- the relative amount of product decreases for a reaction that is endothermic in the forward direction
- the relative amount of product increases for a reaction that is exothermic in the forward direction.

Increase in pressure

Equilibrium shifts to **right**
More B is converted to A

- If pressure decreases, the system will increase the rate of reaction which has the most amount of gas.

Decrease in pressure

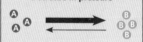

Equilibrium shifts to **left**
More B is made

- If you increase the concentration of a **product**, the system will oppose the change by increasing the rate of the backwards reaction. This lowers the concentration of the products and increases the concentration of the **reactants**.

Increase in product

Equilibrium shifts to **left**
B is converted to A

> Changing pressure on an equilibrium system with no substances in the gas phase will have no effect on the rate of reaction or position of equilibrium.

The effect of changing conditions on reversible reactions

Le Chatelier's Principle

1 What can Le Chatelier's Principle be used for?

....

....

....

2 What does Le Chatelier's Principle state?

....

....

Changing conditions

3 Complete the sentences by filling in the gaps.

The yield .... when more products are added to a reversible reaction

at equilibrium. The yield .... when more reactants are added to a

reversible reaction at equilibrium.

The yield of a product .... if the temperature is increased for a

reversible reaction at equilibrium, whose forward reaction is exothermic. The yield of a product

.... if the temperature is increased for a reversible reaction at

equilibrium, whose forward reaction is endothermic.

4 a) What is the effect on the position of equilibrium if the temperature is increased for a reversible reaction at equilibrium?

....

b) What is the effect on the position of equilibrium if the temperature is decreased for a reversible reaction at equilibrium?

....

5 a) What is the effect on the position of equilibrium if the pressure is increased for a gas phase reversible reaction at equilibrium?

....

b) What is the effect on the position of equilibrium if the pressure is decreased for a gas phase reversible reaction at equilibrium?

....

c) What is the effect on the position of equilibrium if the pressure is increased for a reversible reaction at equilibrium with no substances that are gases?

....

7 Crude oil

Fractional distillation

Crude oil is found in rocks as the remains of an ancient **biomass** consisting mainly of plankton that was buried in mud. It is a:

- **mixture** of **hydrocarbons**
- **fossil fuel**
- **finite** resource.

Fractional distillation is used to separate crude oil into useful fractions. Each fraction is a mixture of **hydrocarbons** with similar **boiling points**.

- It can be done in a lab – a thermometer reads the temperature of the boiling point of the fraction that is being collected. When the temperature shoots up, the collecting flask is changed and the next fraction is collected. This is repeated until all the fractions have been collected.
- In industry, the crude oil is pumped into a furnace and evaporates. The vapour enters the **fractionating column** and the different fractions condense at different heights.

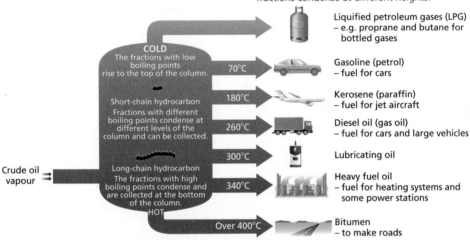

COLD
The fractions with low boiling points rise to the top of the column.

Short-chain hydrocarbon
Fractions with different boiling points condense at different levels of the column and can be collected.

Long-chain hydrocarbon
The fractions with high boiling points condense and are collected at the bottom of the column.
HOT

Crude oil vapour

70°C — Liquified petroleum gases (LPG) – e.g. proprane and butane for bottled gases

70°C — Gasoline (petrol) – fuel for cars

180°C — Kerosene (paraffin) – fuel for jet aircraft

260°C — Diesel oil (gas oil) – fuel for cars and large vehicles

300°C — Lubricating oil

340°C — Heavy fuel oil – fuel for heating systems and some power stations

Over 400°C — Bitumen – to make roads

The fractions of crude oil are processed by the petrochemical industry to make fuels and materials like solvents, lubricants, polymers and detergents.

Hydrocarbons

Hydrocarbons contain only hydrogen and carbon atoms. The bigger the hydrocarbon:

- the higher its boiling point
- the more **viscous** it is
- the less easily **flammable** it is.

Hydrocarbons are often used as fuels and are combusted when used. When **complete combustion** happens, the atoms are fully oxidised and only water and carbon dioxide are produced.

The general equation for this reaction is:

hydrocarbon + oxygen → carbon dioxide + water

Most hydrocarbons in crude oil are **alkanes** with the **general formula** C_nH_{2n+2}.

Alkanes are:

- a **homologous series**
- **saturated** (contain only single covalent bonds)
- hydrocarbons.

H–C–H with H top, bottom, and sides. The simplest alkane, **methane, CH_4**, is made up of 4 hydrogen atoms and 1 carbon atom.	H–C–C–H structure. **Ethane, C_2H_6** A molecule made up of 2 carbon atoms and 6 hydrogen atoms.	H–C–C–C–H structure. **Propane, C_3H_8** A molecule made up of 3 carbon atoms and 8 hydrogen atoms.	H–C–C–C–C–H structure. **Butane, C_4H_{10}** A molecule made up of 4 carbon atoms and 10 hydrogen atoms.

7 Crude oil

Fractional distillation

1 **a)** What is crude oil made from?

....

b) Where is crude oil found?

....

2 Describe a fraction of crude oil.

....

....

3 Name the **two** physical changes that happen to crude oil to separate it using fractional distillation.

....

Hydrocarbons

4 Which elements are found in hydrocarbons? Circle the correct answers.

Hydrogen	Helium	Oxygen	Carbon	Calcium

5 How does the size of the molecule affect flammability?

....

....

6 Describe what happens in complete combustion.

....

7 What type of bonds are found in alkanes?

....

8 Tick the correct equation for the complete combustion of a hydrocarbon.

hydrocarbon + carbon dioxide ➜ oxygen + water ☐

hydrocarbon + oxygen ➜ carbon dioxide + water ☐

hydrocarbon + nitrogen ➜ carbon monoxide + water ☐

hydrocarbon + carbon ➜ carbon dioxide + water ☐

7) Cracking

Cracking

Cracking is the breaking down of long-chain hydrocarbons to make smaller **hydrocarbons**. This is an **endothermic decomposition** reaction.

The products of cracking are a mixture of:
- **alkanes** – useful for fuels
- **alkenes** – useful for the chemical industry to make products like **polymers**.

Cracking can be done on a small scale in the school lab. The long-chain hydrocarbons are heated until they vapourise and are passed over a catalyst which causes the chemical reaction.

There are two main industrial methods used for cracking:

- **Catalytic cracking** – with a catalyst (e.g. zeolite, aluminium oxide and/or silicon dioxide) and temperatures of 550°C. Very useful for making petrol.
- **Steam cracking** – no catalysts, uses steam, high pressure and temperatures of over 800°C. Makes more alkenes, which are used in the petrochemical industry to make polymers.

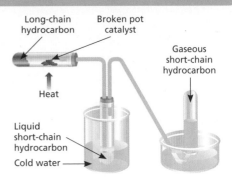

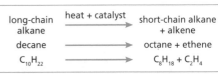

long-chain alkane	heat + catalyst	short-chain alkane + alkene
decane	→	octane + ethene
$C_{10}H_{22}$	→	$C_8H_{18} + C_2H_4$

When crude oil is separated into fractions, there are more long-chain hydrocarbons than are needed. Cracking makes the long-chain hydrocarbons into smaller, more useful hydrocarbons.

Alkenes

Cracking produces a different type of organic molecule with the **general formula** C_nH_{2n}.

Alkenes:
- are a **homologous series**
- are **unsaturated** (contain two fewer hydrogen atoms than the alkane, with the same number of carbon atoms)
- have the **functional group** C=C
- are **hydrocarbons**.

Alkanes and alkenes can be distinguished using a simple laboratory test using bromine water (Br_2(aq)):
- alkanes = no colour change
- alkenes = colour change from orange / brown to colourless.

Alkene	Ethene, C_2H_4	Propene, C_3H_6	Butene, C_4H_8	Pentene, C_5H_{10}
Displayed formula	H\C=C/H with H, H	H\C=C-C-H with H, H, H, H	H\C=C-C-C-H with H, H, H, H, H	H-C=C-C-C-H with H, H, H, H, H, H, H

7 Cracking

Cracking

1 Define the term cracking.

2 What is made during cracking?

3 What happens to the long-chain hydrocarbon when it is cracked in the lab?

4 **a)** What are the conditions for catalytic cracking?

b) What is catalytic cracking mainly used to make?

5 **a)** What are the conditions for steam cracking?

b) What is steam cracking mainly used to make?

Alkenes

6 **a)** Why are alkenes classified as hydrocarbons?

b) Why are alkenes described as unsaturated?

c) What is the functional group in alkenes?

7 Which chemical can be used to distinguish between alkanes and alkenes?

8 What is the molecular formula of propene?

Reactions of hydrocarbons

Hydrocarbons contain only carbon and hydrogen atoms. They can be oxidised by:
- complete combustion – excess oxygen and only carbon dioxide and water are made
- incomplete combustion – limited oxygen and a mixture of carbon (soot), carbon dioxide and carbon monoxide are made.

Alkenes are more reactive than alkanes because they have the **functional group** C=C. They undergo:
- an incomplete combustion with a smoky flame
- an **addition reaction** with hydrogen to become a **saturated alkane**:

Ethene   Hydrogen   Ethane

- an addition reaction with water to become an **alcohol**:

Ethene   Steam   Ethanol

- an addition reaction with a **halogen** to make a **haloalkane**:

Ethene   Chlorine   Haloalkane

Alcohols and carboxylic acids

Alcohols are organic molecules that contain carbon, hydrogen and oxygen atoms. Alcohols:
- are a **homologous series** (family of chemicals)
- have the functional group -OH.

Methanol, CH_3OH	
Ethanol, CH_3CH_2OH	
Propanol, $CH_3CH_2CH_2OH$	
Butanol, $CH_3CH_2CH_2CH_2OH$	

Aqueous solutions of ethanol can be made from **fermentation** using yeast and temperatures of about 40°C. The chemical equations for this reaction are:

glucose → carbon dioxide + ethanol

$C_6H_{12}O_6 \rightarrow 2C_2H_5OH + 2CO_2$.

Alcohols can undergo:
- complete combustion to make carbon dioxide and water: $CH_3CH_2OH + 3O_2 \rightarrow 3H_2O + 2CO_2$
- incomplete combustion to make a mixture of carbon, carbon dioxide and carbon monoxide
- a reaction with sodium to produce a **salt** and hydrogen.

Carboxylic acids are organic molecules that contain carbon, hydrogen and oxygen atoms. They:
- are a homologous series
- have the functional group -COOH
- are **weak acids** (partially ionise in solution).

Methanoic acid, HCOOH	
Ethanoic acid, CH_3COOH	
Propanoic acid, C_2H_5COOH	
Butanoic acid, C_3H_7COOH	

Carboxylic acids can undergo:
- a **neutralisation** reaction with carbonates to make carbon dioxide, a salt and water
- a reaction with alcohols to make an ester and water:

Ethanol

Ethanoic acid

Ethyl ethanoate

Water

RETRIEVE

7 Homologous series

Reactions of hydrocarbons

1 a) What are the products of complete combustion of a hydrocarbon?

b) What are the products of incomplete combustion of a hydrocarbon?

2 What type of chemical reaction happens when bromine reacts with ethene?

3 What is the product when butene reacts with steam?

Alcohols and carboxylic acids

4 a) What are the products of complete combustion of an alcohol?

b) What are the products of incomplete combustion of an alcohol?

5 What is the functional group of an alcohol? Tick the correct answer.

-OH ☐   -COOH ☐

-CO ☐   -CH ☐

6 What is made when sodium reacts with an alcohol?

7 What is the functional group of a carboxylic acid? Tick the correct answer.

-OH ☐   -COOH ☐

-CO ☐   -CH ☐

8 HT Why is a carboxylic acid a weak acid?

9 What is made when an alcohol reacts with a carboxylic acid?

7 Polymers

Polymers

Polymers are very long molecules made from small repeating units called **monomers**. The **atoms** are held together by **covalent bonds**. Polymers are made in a chemical reaction called **polymerisation**.

Addition polymers

Addition polymers are made from:
- **Addition reactions** – 100% **atom economy** and makes only one **product**, the polymer.
- **Alkenes (monomers)** – the reactive C=C **functional group** which opens up to link all the molecules.

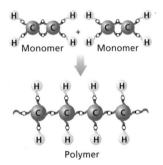

Monomer + Monomer

Polymer

Ethene molecules are monomers with the C=C functional group, which react together to make polyethene. This can be shown with a chemical equation where n means lots of and the **repeating unit** is shown to represent the polymer:

$$\begin{matrix} H & & H \\ & C=C & \\ H & & H \end{matrix} \rightarrow \left(\begin{matrix} H & H \\ C-C \\ H & H \end{matrix} \right)_n$$

Propene molecules react to make polypropene:

$$n\ \begin{matrix} H & & CH_3 \\ & C=C & \\ H & & H \end{matrix} \rightarrow \left(\begin{matrix} H & CH_3 \\ C-C \\ H & H \end{matrix} \right)_n$$

> Addition polymers have strong covalent bonds which do not break easily. This is what makes them non-biodegradable.

(HT) Condensation polymers

Condensation polymers are made from:
- **condensation reactions**, where the polymer and a small molecule are made
- **monomers** with two functional groups.

Polyesters are made from the reaction between a **diol** (an organic molecule with two **alcohol** groups) and a **dicarboxylic acid** (an organic molecule with two carboxylic acid groups). Each monomer is linked together by an **ester** group. A **block diagram** can show a simplified model of this reaction:

$$n\ \begin{matrix} H & H \\ H-O-C-C-O-H \\ H & H \end{matrix} + n\ \begin{matrix} O & H & H & H & H & O \\ C-C-C-C-C-C \\ H-O & H & H & H & H & O-H \end{matrix}$$

Often simplified to:

HO—☐—OH   HOOC—☐—COOH

↓

$$\left[\begin{matrix} H & H & O & H & H & H & H & O \\ C-C-O & C-C-C-C-C \\ H & H & & H & H & H & H & O \end{matrix} \right]_n$$

Often simplified to:

$$\left(O-☐-O-CO-☐-CO \right)_n$$

$$+\ 2H_2O$$

> Condensation polymers can chemically react with water, and this can break the bonds between the two different monomers. So, condensation polymers may be biodegradable.

(7) Polymers

Polymers

1 a) What is a polymer?

b) What is a monomer?

c) What is polymerisation?

Addition polymers

2 What type of monomer is used to make addition polymers? Tick the correct answer.

Alkanes ☐   Alkenes ☐

Esters ☐   Alcohols ☐

3 How many types of monomers are used to make an addition polymer? Tick the correct answer.

1 ☐   2 ☐   5 ☐   10 ☐

4 Why does addition polymerisation have 100% atom economy?

5 What is the name of the monomer which makes polybutene?

6 What is the name of the polymer which is made from pentene?

Condensation polymers

7 How many types of monomers are used to make a condensation polymer? Tick the correct answer.

1 ☐   2 ☐   5 ☐   10 ☐

8 What is made during condensation polymerisation?

9 What are the names of the monomers used to make polyesters?

Natural polymers

Polymers are long-chain **molecules**. They can be found naturally in living systems. Examples are polypeptides and polysaccharides.

Polypeptides, like **protein**, are made from amino acids and are used to make **enzymes**, **hormones** and muscles.	Amino acids Peptide Protein
Polysaccharides include: • **starch** – made from **glucose** molecules and used as an energy store in plants • **cellulose** – made from glucose molecules and used to make strong cell walls in plant cells.	Starch Cellulose Amylose Amylopectin

Proteins are natural polymers made from monomers which are also **amino acids**. Different amino acids can be combined in the same chain to produce proteins.

(HT) Amino acids

Amino acids:
• are **organic molecules** that contain carbon, hydrogen, nitrogen and oxygen atoms
• have two different **functional groups** (-NH_2 and -COOH)
• can undergo **condensation polymerisation**.

Glycine is an amino acid with the **structural formula**, H_2NCH_2COOH. The **amino group** on one molecule can react with the **carboxylic acid**

on the other molecule to make a **peptide bond** linking the two amino acids:

This can happen many times to make a polymer. This is a **condensation reaction** and produces a polypeptide and a molecule of water.

DNA

Deoxyribonucleic acid (DNA):
• is a large molecule essential for life
• encodes genetic instructions for the development and functioning of living organisms and viruses.

Most DNA molecules are two polymer chains, made from four different monomers called **nucleotides**, in the form of a double helix. This is found in the nucleus of all **eukaryotic cells**.

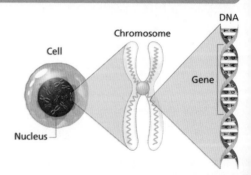

Cell
Nucleus
Chromosome
Gene
DNA

7 Natural polymers

Natural polymers

1 What is the monomer used to make starch?

....

2 What is the monomer used to make cellulose?

....

Amino acids

3 How many functional groups does an amino acid molecule have? Tick the correct answer.

1 ☐   2 ☐   3 ☐   4 ☐

4 **HT** Complete the diagram by adding the labels shown in the box in the correct places.

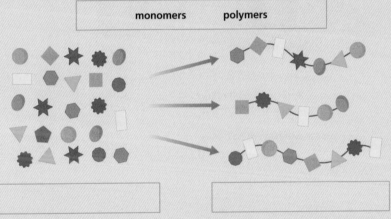

monomers   polymers

5 What is the name of the polymer formed from amino acids?

....

DNA

6 Where is DNA found?

....

7 What is the purpose of DNA?

....

8 Name the monomers that make up DNA.

....

9 Describe the shape of DNA.

....

⑧ Pure substances

Everyday pure substances and chemically pure substances

In everyday language substances like milk and water can be described as **pure**. In this context we mean that milk is in its natural state with nothing added, and water is in its natural state with nothing added (**unadulterated**).

In chemistry, milk is not a pure substance as it is a mixture of substances. Water is not a pure substance either, as it is a mixture of substances. In chemistry there is a specific meaning for a pure substance:

- a single **element** or **compound**
- not mixed with any other substance.

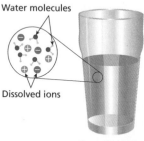

Water molecules

Dissolved ions

Everyday water

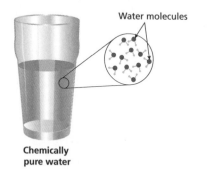

Water molecules

Chemically pure water

Pure substances have specific **melting** and **boiling points**, whereas mixtures will change state over a range of temperatures.

Every pure substance has a unique and precise melting and boiling point, which can be used to identify it. For example, a colourless liquid with a melting point of 0°C and a boiling point of 100°C would be pure water; a liquid with a melting point of –114°C and a boiling point of 78°C would be pure ethanol.

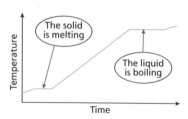

The solid is melting

The liquid is boiling

Temperature–time graph for heating a substance

Formulations

A formulation is:
- a **mixture**
- not pure
- designed to be a useful product.

In a formulation, the **components** of the mixture are chosen for a particular purpose. These components are carefully measured and mixed together to make a product with the desired properties.

Use of formulations include:
- fuels
- cleaning agents
- paints
- medicines
- alloys
- foods
- fertilisers.

⑧ Pure substances

Everyday pure substances and chemically pure substances

1 What does pure mean in everyday language?

....

2 What does pure mean in terms of chemistry?

....

3 What particle(s) are found in chemically pure water?

....

4 Explain why milk is not chemically pure.

....

....

5 What substances can be chemically pure?

....

....

Formulations

6 **a)** What is a formulation?

....

b) Why is a formulation not chemically pure?

....

7 Give **one** use of formulations.

....

8 Why is a formulation also a type of mixture?

....

....

9 How are components of a formulation chosen?

....

....

⑧ Chromatography

Chromatography

Chromatography is a separating technique used to:
- determine if a substance is pure or impure
- separate mixtures
- identify substances.

The process of chromatography is as follows:
- The substance which is being analysed – like an ink or a dye – is put on absorbent paper. This is the **stationary phase** as the paper doesn't move.
- The edge of the paper is placed into a solvent that moves up through the paper. This is the **mobile phase**.
- The different parts of the mixture are attracted to the paper and solvent by different amounts and this causes them to separate.

Paper chromatography

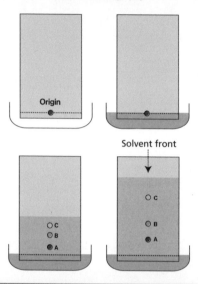

Interpreting a chromatogram

One spot on a **chromatogram** shows that a substance is pure. Multiple dots on a chromatogram show how many components there are in the mixture. The **retention factor, R_f,** value can be used to determine the identity of the substance.

The R_f value is:
- the **ratio** of the distance moved by a substance (centre of spot from origin) to the distance moved by the **solvent**
- different when different solvents (mobile phases) are used
- used to determine an unknown substance.

This chromatogram shows a pure substance as there is only one dot in the chromatogram.

The R_f value can calculated by:

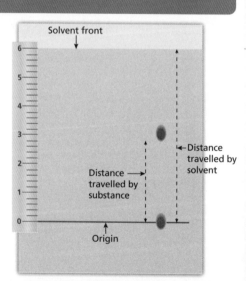

$$R_f \text{ value} = \frac{\text{distance travelled by substance}}{\text{distance travelled by solvent}} = \frac{3}{6} = 0.5$$

Substances can be identified using reference samples on the same chromatogram. The unknown and known substances are used to create the same chromatogram. Any dots that are in line with each other are the same substance.

8 Chromatography

Chromatography

1 The diagram shows the process of chromatography.

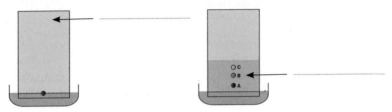

a) Label the mobile phase on the diagram.

b) Label the stationary phase on the diagram.

2 Explain how chromatography works.

....

....

....

Interpreting a chromatogram

3 What is a chromatogram?

....

4 Complete the sentences by filling in the gaps.

If only one dot appears on the chromatogram, the substance is ....

If more than one dot appears on the chromatogram, the substance is

.... or a ....

5 What is R_f?

....

....

6 What is the solvent front?

....

....

7 a) How do you calculate R_f?

....

b) What information does the R_f value give you?

....

⑧ Identifying common gases

Producing gas

In a **chemical reaction**, if a gas is produced you may notice:
- **effervescence** (seeing bubbles and hearing fizzing)
- the **mass** decreases as gas is lost to the atmosphere.

Gases can be collected in different ways:

Gases that do not easily dissolve in water, like hydrogen, can be collected by **displacement** of water.	Gases that are denser than air, like carbon dioxide, can be collected by **downward delivery**.	Gases that are less dense than air, like hydrogen, can be collected by **upward delivery**.
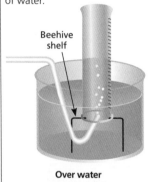 Beehive shelf **Over water**	 **Downward delivery (upward displacement of air)**	 **Upward delivery (downward displacement of air)**

All gases can be collected and measured using a gas syringe.

Gas tests

If a gas is hydrogen, $H_2(g)$: • Test: use a lighted splint • Result: hear a squeaky pop  Pop	If a gas is carbon dioxide, $CO_2(g)$: • Test: add limewater to the gas and shake • Result: limewater changes from colourless to cloudy CO_2 gas Limewater 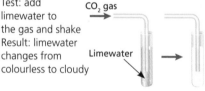
If a gas is oxygen, $O_2(g)$: • Test: use a glowing splint • Result: relights 	If a gas is chlorine, $Cl_2(g)$: • Test: put a piece of damp litmus paper into the gas • Result: turns white (bleaches)

8 Identifying common gases

Producing gas

1 What **two** observations are described by the term effervescence?

_____

2 Explain why mass might appear to drop when a gas is made in a chemical reaction.

_____

_____

3 Draw lines to link the method of gas collection with when it would be used.

Method of gas collection	**When it would be used**
Displacement of water	When the gas is denser than air
Downward delivery	When the gas is less dense than air
Upward delivery	When the gas has low solubility in water

4 **a)** Give an example of a gas that can be collected by displacement of water.

_____

b) Give an example of a gas that can be collected by downward delivery.

_____

c) Give an example of a gas that can be collected by upward delivery.

_____

Gas tests

5 How do you test for hydrogen gas?

_____

6 What happens to a glowing splint if oxygen is present?

_____

7 Which substance is used to test for carbon dioxide?

_____

8 What indicator is used to test for chlorine?

_____

9 What happens to indicator paper when it is put into chlorine gas?

_____

(8) Identifying positive ions

Flame test

Metals make **cations** (positive ions) by:
- losing their outer shell electrons
- becoming positive ions.

Some metal ions cause characteristic flame colours, which can be used to identify them. This property can be used in an analytical test called a **flame test**:
- Put a small sample of the substance on a clean loop of nichrome wire.
- Put the loop into the blue Bunsen flame.
- Use the colour to determine which metal ion is present.

The colours are:
- green / blue for **copper**
- brick red for **calcium**
- red for **lithium**
- lilac for **potassium**
- yellow / orange for **sodium**.

> Flame colours can be masked and unable to interpreted if there are a mixture of metal ions in the sample.

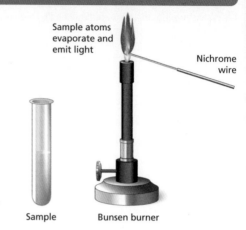

Sample atoms evaporate and emit light

Nichrome wire

Sample   Bunsen burner

Ba   Ca   Li   Na   Cu   K

Precipitation test

Precipitation reactions are when two solutions react together to make at least one **insoluble** solid and the **solution** becomes **turbid** (cloudy).

Sodium hydroxide solution can react with some metal ions in a precipitation reaction. The colour of the precipitate is used to identify the metal ion:

- Aluminium ions (Al^{3+}) form a white precipitate that dissolves in excess sodium hydroxide.

- Calcium ions (Ca^{2+}) and magnesium ions (Mg^{2+}) form white precipitates. The white precipitate remains (doesn't dissolve) even in excess sodium hydroxide:

> HT – **Ionic equations**: $Ca^{2+} + 2OH^- \rightarrow Ca(OH)_2$
> and $Mg^{2+} + 2OH^- \rightarrow Mg(OH)_2$

- Copper(II) (Cu^{2+}) forms a blue precipitate.
  - Example **balanced symbol equation**:
    $$CuSO_4(aq) + 2NaOH(aq) \rightarrow Na_2SO_4(aq) + Cu(OH)_2(s)$$

> HT – Ionic equation: $Cu^{2+} + 2OH^- \rightarrow Cu(OH)_2$

- Iron(II) (Fe^{2+}) forms a green precipitate.
  - Example balanced symbol equation:
    $$FeSO_4(aq) + 2NaOH(aq) \rightarrow Na_2SO_4(aq) + Fe(OH)_2(s)$$

> HT – Ionic equation: $Fe^{2+} + 2OH^- \rightarrow Fe(OH)_2$

- Iron(III) ions, (Fe^{3+}) forms a brown precipitate.
  - Example balanced symbol equation:
    $$Fe_2(SO_4)_3(aq) + 6NaOH(aq) \rightarrow 3Na_2SO_4(aq) + 2Fe(OH)_3(s)$$

> HT – Ionic equation: $Fe^{3+} + 3OH^- \rightarrow Fe(OH)_3$

8 Identifying positive ions

Flame test

1 Why can flame tests be used to identify some metal ions?

....

....

2 Draw lines to link the colour of the flame during a flame test, to the metal ions present.

Colour of flame	**Metal ions present**
Yellow / orange flame	Potassium
Lilac flame	Calcium
Brick red flame	Sodium

Precipitation test

3 What is a precipitate?

....

....

4 What colour is the precipitate when magnesium ions react with sodium hydroxide?

....

5 How do you tell the difference from two samples of magnesium and aluminium ions?

....

....

6 a) What is the formula of the ion which makes a blue precipitate with sodium hydroxide solution?

....

b) What colour is the precipitate when iron(II) ions form a precipitate with sodium hydroxide solution?

....

7 Write a balanced symbol equation for the reaction between copper(II) sulfate and sodium hydroxide solution.

....

....

8 Identifying negative ions

Carbonates

Carbonates (CO_3^{2-}) are:
- **ionic compounds**
- **basic** (react with acids)
- negative **ions** (**anion**).

The simple laboratory test for a carbonate involves adding some dilute acid, e.g. 0.5 mol/dm³ hydrochloric acid to the sample.

> HT • $CO_3^{2-}(s) + 2H^+(aq) \rightarrow CO_2(g) + H_2O(l)$
> • $CO_3^{2-}(aq) + 2H^+(aq) \rightarrow CO_2(g) + H_2O(l)$

If **effervescence** (fizzing and bubbles) happens, collect and test the gas with limewater. If the limewater turns from colourless to cloudy then carbon dioxide is the gas and the original sample was a carbonate.

Test for a carbonate

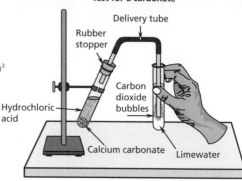

Adding calcium carbonate to hydrochloric acid will produce calcium chloride, carbon dioxide and water.
$$CaCO_3 + 2HCl \longrightarrow CaCl_2 + CO_2 + H_2O$$

Halides test

Halides:
- are negative ions (anions)
- have a 1- charge
- are made from a **halogen atom** that has gained an **electron** into its outer shell.

Halides produce coloured **precipitates** with silver nitrate solution in the presence of nitric acid. The colour of the precipitate can be used to infer the halide present:
- Chloride makes a white precipitate of silver chloride: HT $Cl^-(aq) + Ag^+(aq) \rightarrow AgCl(s)$
- Bromide makes a pale cream precipitate of silver bromide: HT $Br^-(aq) + Ag^+(aq) \rightarrow AgBr(s)$
- Iodide makes a pale yellow precipitate of silver iodide: HT $I^-(aq) + Ag^+(aq) \rightarrow AgI(s)$

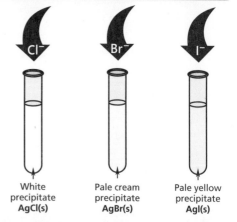

White precipitate **AgCl(s)**

Pale cream precipitate **AgBr(s)**

Pale yellow precipitate **AgI(s)**

Sulfates test

Sulfates (SO_4^{2-}):
- are positive ions (cations)
- have a 2- charge.

Sulfates produce a white precipitate with barium chloride solution and hydrochloric acid:

> HT $Ba^{2+}(aq) + SO_4^{2-}(aq) \rightarrow BaSO_4(s)$

It is important to choose the correct acid for each test as:
- sulfuric acid cannot be used in the sulfate test because it contains the sulfate ion.
- hydrochloric acid cannot be used in the halide test because it contains the chloride ion.

In the simple laboratory test for halide ions and sulfate ions, dilute acid is added to the test solution. This acidifies the solution and reacts with any dissolved carbon dioxide or carbonate ions in the test sample. These ions would cause a white precipitate and this result would cause an incorrect conclusion.

⑧ Identifying negative ions

Carbonates

1 What is the formula of a carbonate ion?

....

2 What do you observe when a dilute acid is added to a carbonate?

....

3 What happens to limewater when carbon dioxide is blown through it?

....

4 What is an anion?

....

Halides test

5 What are the substances used in the halide test?

....

6 What colour is the precipitate made from chloride (Cl^-) ions?

....

7 **HT** Write a balanced ionic equation for the precipitation reaction of bromide ions and acidified silver nitrate solution.

....

8 Which halide makes a yellow precipitate? Tick the correct answer.

Iodine ☐

Oxide ☐

Iodide ☐

Sulfide ☐

Sulfates test

9 What are the substances used in the sulfate test?

....

10 **HT** Write a balanced ionic equation for a positive result of the sulfate test.

....

8 Instrumental methods

Instrumental methods

As technology has developed, chemical **analysis** of **elements** and **compounds** has changed to involve machines. The hands-on analysis has been replaced by **instrumental methods** which are:
- more **accurate** – give a value closer to the true value
- more **sensitive** – give the correct positive result for smaller amounts of a substance
- faster.

However, instrumental methods usually require more expensive equipment and specialist training to use them and to interpret results.

Flame emission spectroscopy

Metal **ions** can be identified using the instrumental technique called **flame emission spectroscopy**.
- The metal ion solution is sprayed into a flame.
- Light given out by the flame is passed through a spectroscope or spectrometer (a machine that measures the light given out from a sample).
- A **spectrum** is produced with lines corresponding to the energy released by the ions in the flame.

Each metal ion will produce a characteristic **emissions spectrum**, which looks like a coloured bar code. This unique pattern can be used to identify the metal ions present in the sample – even in a sample containing a mixture of metal ions.

Different concentrations of a metal ion solution can be put into the flame spectroscope and a reading taken for each. These data can be used to plot a calibration curve to measure the concentration of an unknown solution of that metal ion.

Flame emission spectroscopy

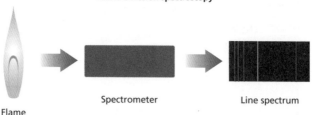

Flame — Spectrometer — Line spectrum

Comparing methods of identifying metal ions

Flame tests and flame emission spectroscopy can both identify metal ions in a sample quickly. Each has its benefits and drawbacks, and it is important that you can justify the use of one method over the other.
- Flame tests use simpler equipment and are a cheaper way to analyse. But the colours are subjective (can appear different for each scientist) leading to some **uncertainty** in the results.
- Flame emission spectroscopy can identify very **dilute** solutions of the ions and give more information (a wider variety of ions, determine metal ions in a mixture, as well as measure the concentrations of ions). But the equipment is expensive and specialist training is needed to use the equipment and to interpret the results.

⑧ Instrumental methods

Instrumental methods

1 What is an instrumental method?

_____

_____

2 Suggest why instrumental analysis has developed.

_____

3 **a)** What are the advantages of instrumental methods over traditional hands-on analysis?

_____

b) What are the drawbacks of instrumental methods over traditional hands-on analysis?

_____

Flame emission spectroscopy

4 What can be identified from using flame emission spectroscopy?

_____

5 What is produced by flame emission spectroscopy?

_____

6 What is a spectroscope?

_____

_____

Comparing methods of identifying metal ions

7 What hands-on analytical technique has been replaced by flame emission spectroscopy?

_____

8 **a)** Describe the benefits of flame emission spectroscopy over flame tests.

_____

_____

b) What are the benefits of a flame test over flame emission spectroscopy?

_____

_____

_____

The development of the Earth's atmosphere

Air and the development of the atmosphere

The envelope of gas around our planet is the **atmosphere**. **Air** is the **mixture** of gases found in the atmosphere and has been the same for about the last 200 million years. Dry air is made of:

- 80% ($\frac{4}{5}$) nitrogen, N_2
- 20% ($\frac{1}{5}$) oxygen, O_2
- trace amounts of carbon dioxide, water vapour and **noble gases**.

Evidence for the development of the atmosphere is limited because of the huge timescale. We are reliant on evidence from rocks, ice cores and space exploration to make hypotheses and plausible theories.

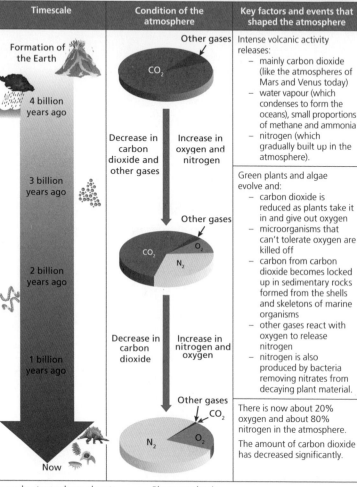

Timescale	Condition of the atmosphere	Key factors and events that shaped the atmosphere
Formation of the Earth 4 billion years ago	Other gases CO_2	Intense volcanic activity releases: – mainly carbon dioxide (like the atmospheres of Mars and Venus today) – water vapour (which condenses to form the oceans), small proportions of methane and ammonia – nitrogen (which gradually built up in the atmosphere).
3 billion years ago	Decrease in carbon dioxide and other gases Increase in oxygen and nitrogen	Green plants and algae evolve and: – carbon dioxide is reduced as plants take it in and give out oxygen – microorganisms that can't tolerate oxygen are killed off – carbon from carbon dioxide becomes locked up in sedimentary rocks formed from the shells and skeletons of marine organisms – other gases react with oxygen to release nitrogen – nitrogen is also produced by bacteria removing nitrates from decaying plant material.
2 billion years ago	Other gases CO_2 O_2 N_2	
1 billion years ago	Decrease in carbon dioxide Increase in nitrogen and oxygen	There is now about 20% oxygen and about 80% nitrogen in the atmosphere. The amount of carbon dioxide has decreased significantly.
Now	Other gases CO_2 N_2 O_2	

Oxygen appeared in the early atmosphere about 2.7 billion years ago when **algae** developed. These organisms, later joined by green plants, use **photosynthesis** to obtain the energy from the sun that they needed for life:

$$\text{carbon dioxide} + \text{water} \rightarrow \text{glucose} + \text{oxygen}$$
$$6CO_2 + 6H_2O \rightarrow C_6H_{12}O_6 + 6O_2$$

The oxygen in the atmosphere increased over the next billion years until animal life could evolve.

Levels of carbon dioxide in the early atmosphere reduced because of:

- Photosynthesis.
- The formation of **carbonate rocks** like limestone. Carbon dioxide dissolved into ocean water and made **soluble carbonates**. These **precipitate** out to make sedimentary rocks, which store carbon dioxide for a very long time.
- The formation of **fossil fuels**, made from ancient **biomass** locked underground, which chemically changes into **hydrocarbons** and locks away some of the carbon from the atmosphere. The stored carbon is then released when the fossil fuels are **combusted** as they are used.

The development of the Earth's atmosphere

Air and the development of the atmosphere

1 What is the atmosphere?

2 **a)** Which gas is the main component of dry air?

b) What (approximate) percentage of dry air is oxygen?

c) What (approximate) fraction of dry air is nitrogen?

3 Which planets in our solar system have a similar atmosphere today as early Earth had?

4 Where did the gases in the early atmosphere come from?

5 Which organism produced the first oxygen in the atmosphere?

6 What is the name of the process that made oxygen in the atmosphere?

7 What type of rock is limestone?

8 Give **three** ways that carbon dioxide was reduced from the early atmosphere.

Causes of global climate change

Greenhouse gases and the greenhouse effect

Gases in the Earth's **atmosphere** absorb reflected **long-wave radiation** from the Earth's surface. They trap heat energy and keep the average world temperature 14–16°C. These **greenhouse gases** include water vapour, $H_2O(g)$; carbon dioxide, $CO_2(g)$ and methane, $CH_4(g)$.

The **greenhouse effect** is the natural process in the atmosphere that keeps Earth's average temperature high enough to support life.

Since the industrial revolution, where **combustion** of **fossil fuels** was large scale, the proportion of greenhouse gases has been increasing, which has led to **global warming** – a rise in the average temperature of Earth beyond the natural level.

This is caused by:
- carbon dioxide increasing through combustion of fossil fuels

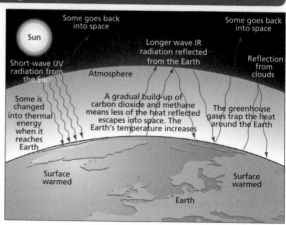

- methane increasing by animal farming, rice paddies, landfills and thawing of permafrost.

Evidence is collected by scientists, which is **peer reviewed** and has been accepted as **accurate**. Scientists have **concluded** that increasing the concentration of greenhouse gases in the atmosphere is causing global warming and leading to **climate change**.

Global climate change and carbon footprint

The effects of climate change include:
- changes in weather patterns – increased flooding, drought and water shortages in summer
- an increase in natural disasters – more powerful and frequent tropical storms, increased number and severity of heatwaves and wildfires
- melting of ice caps – increased sea levels and flooding of low-lying land
- changes to habitats – reduction in **biodiversity**, and industries will need to change (e.g. no snow in ski resorts, different crops needing to be grown by farmers)
- an increase in some diseases – malaria risk in more countries.

Adapting to the new climate may include:
- relocation of people
- flood defence schemes
- developing new technologies including low carbon fuels, more energy efficient devices and bioengineering crops to withstand extreme weather
- reduction of greenhouse gases by using carbon capture technologies and planting forests

- education of the population
- international treaties.

The total amount of greenhouse gases over the full life cycle of a product, service or event is the **carbon footprint**. This is measured in carbon dioxide equivalent (CO_2e). It can be reduced by:
- extending the lifespan of the product by reusing or recycling
- using renewable energy resources for manufacturing and transportation
- reducing energy used by using local resources and lightweight packaging
- reducing energy loss by insulating or adding lubricants to reduce friction in machines.

It is difficult for people to change their habits but this can be encouraged by laws and financial incentives from government.

It is difficult to make an accurate model of climate change as simplifications must be used. The media use opinion and speculation as well as the outcome of models, which may lead to bias.

9 Causes of global climate change

Greenhouse gases and the greenhouse effect

1 Name the **three** main greenhouse gases in our atmosphere.

2 Why is it important that greenhouse gases are in our atmosphere?

3 Explain what humans are doing to the proportion of greenhouse gases in our atmosphere and how they are doing this.

4 What is global warming?

Global climate change and carbon footprint

5 What is climate change?

6 Explain why sea levels might rise.

7 What is a carbon footprint?

8 What is the unit that carbon footprint is measured in?

⑨ Air pollution

Combustion of fuels

Chemical **fuels** are **combusted** when they are used, and are a major source of atmospheric **pollution**. **Fossil fuels** like coal, oil and natural gas contain **hydrocarbons**, which are the fuel, but also sulfur impurities. When these fuels are combusted they form a **mixture** of **products**:

- water
- carbon dioxide
- carbon monoxide
- carbon (soot)
- sulfur dioxide.

When fuels are combusted in enclosed engines, high pressure and heat can cause nitrogen in the air to **oxidise** to form nitrogen oxides, NOx.

Sulfur dioxide and nitrogen oxides are both **acidic** gases and can cause irritation to the human respiratory system.

Carbon monoxide

Carbon monoxide, CO, is:
- a **toxic** gas
- produced when there is **incomplete combustion** of hydrocarbons
- colourless and odourless.

When carbon monoxide is breathed in, it diffuses into the blood and binds to the haemoglobin in the red blood cells stronger than oxygen. This means that the oxygen-carrying capacity of the blood is reduced. The person will experience light-headedness, headaches and confusion. Carbon monoxide poisoning can lead to death.

Acid rain and global dimming

Rainwater has a **pH** of about 5.5. **Acid rain** is caused when acidic gases dissolve into the rainwater and cause the pH of the rain to be lowered. Acid rain is caused by:
- sulfur dioxide forming sulfuric acid, H_2SO_4
- oxides of nitrogen forming nitric acid, HNO_3.

> NOx are oxides of nitrogen and include nitrogen monoxide, NO, and nitrogen dioxide, NO_2. These are the oxides of nitrogen that are most likely to cause air pollution.

Incomplete combustion of hydrocarbons can cause particles of carbon (soot) to be formed. These are released into the air and cause respiratory problems in humans. In the atmosphere, these particles reflect sunlight and cause **global dimming**.

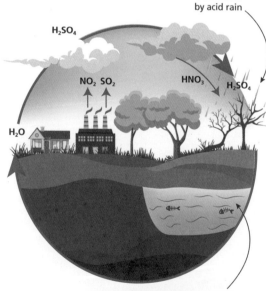

Trees killed by acid rain

H_2SO_4

NO_2 SO_2 HNO_3 H_2SO_4

H_2O

Water becomes acidified causing fish to die

9 Air pollution

Combustion of fuels

1 Which element is an impurity in fuels and can make acidic gases?

....

2 How is NOx made?

....

....

Carbon monoxide

3 **a)** What is the formula of carbon monoxide? Tick the correct answer.

CO ☐   COx ☐

CO_2 ☐ CO_3 ☐

b) What type of combustion causes carbon monoxide to be made?

....

c) How does carbon monoxide affect the human body?

....

....

Acid rain and global dimming

4 What acid can be formed by sulfur dioxide?

....

5 What acid can be formed by nitrogen oxides?

....

6 Define the term acid rain.

....

....

7 Which atmospheric pollutant causes global dimming?

....

8 Complete the following sentence by filling in the gaps.

Global dimming is caused by .... in the atmosphere, which

.... sunlight.

10 Earth's resources

Resources

Resources are everything in our **environment** that are available to help us satisfy our needs and wants. Humans use the Earth's resources to provide warmth, shelter, food and transport.

Resources are **natural** or **synthetic**.

Natural	Synthetic
Natural resources are used chemically unchanged from the Earth to support life and meet people's needs. They include substances like fossil fuels and conditions like sunlight. **Agriculture** works with natural resources to provide food, timber, clothing and fuels.	**Synthetic materials** are made by chemically changing a natural resource to make a new material. For example, rubber is a natural resource that can be vulcanised by reacting it with sulfur to make it more durable and hardwearing, for use as car tyres. Vulcanised rubber is a synthetic material.

Coal resources
Forest resources
Animal resources
Wind and solar resources
Mineral resources
Oil resources
Soil resources
Water resources

Renewable and finite resources

Resources can be classified by their availability:

* **Renewable resources** are resources that can be replaced as they are being used.
* **Finite resources** are being used up faster than the Earth can replace them.

Finite resources can be found in the Earth, oceans and atmosphere. They are processed to provide energy and materials.

Renewable resources

Wind

Hydropower

Solar

Geothermal

Biomass

Finite resources

Oil

Coal

Nuclear

Natural gas

The role of chemistry

Chemistry can improve agriculture and industrial processes to create new products and improve **sustainability**. **Sustainable development** ensures that the needs of the people are met today, whilst ensuring that there are enough resources for future generations too.

10 Earth's resources

Resources

1 What do we use resources for?

2 **a)** What is a natural resource?

b) What is a synthetic resource?

3 **a)** Which classification – natural or synthetic – is metal ore?

b) Which classification – natural or synthetic – is a pure metal?

Renewable and finite resources

4 What is a renewable resource?

5 What is a finite resource?

6 Put a tick in the correct column of the table to say whether each resource is renewable or finite.

Resource	Renewable resource	Finite resource
Biomass		
Nuclear		
Coal		
Geothermal		
Wind		
Oil		

The role of chemistry

7 What is sustainable development?

10 Potable water

Essential water

Water, H_2O, is a **natural resource** which is needed for all life. **Potable water** is water that is safe to drink and has sufficiently low levels of:

- **dissolved salts**
- **microbes**.

There are lots of different ways to find or make potable water but it depends on availability of water supplies and the local conditions. Fresh water is used in the United Kingdom to make potable water. **Fresh water** is rain water so it has low levels of dissolved substances and can collect in the ground and in lakes and rivers.

To make potable water in the UK:

- a suitable source of fresh water is chosen
- fresh water is passed through filter beds to remove **insoluble** solids
- the water is **sterilised** using chlorine (Cl_2) ozone (O_3) or UV light to destroy pathogens and prevent disease.

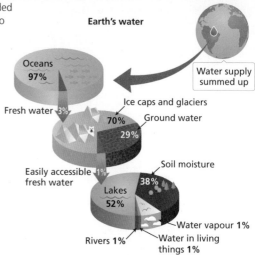

Earth's water

Water supply summed up

Oceans **97%**

Fresh water **3%**

Ice caps and glaciers **70%**

Ground water **29%**

Easily accessible fresh water **1%**

Soil moisture **38%**

Lakes **52%**

Water vapour **1%**

Rivers **1%**

Water in living things **1%**

Pure water contains only water molecules. Pure water is potable but not all potable water is pure water. For example, tap water in the UK has dissolved substances in it but it is still safe to drink.

Desalination

When fresh water supplies are limited, **desalination** of salty water can be used to make potable water. This can be done by **distillation** or **reverse osmosis**.

Distillation	Reverse osmosis
The salty water is boiled and the water vapour condenses into pure water.	The salty water is put under high pressure and passed through a **membrane** which allows the water molecules through, but very few dissolved ions.
This process is expensive as it requires a lot of energy.	This method makes a large amount of waste water, requires expensive membranes and uses a lot of energy.

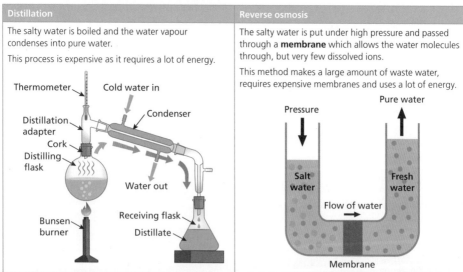

Thermometer | Cold water in

Distillation adapter | Condenser

Cork

Distilling flask

Water out

Bunsen burner

Receiving flask

Distillate

Pressure | Pure water

Salt water | Fresh water

Flow of water

Membrane

10) Potable water

Essential water

1 Describe the difference between pure water and potable water.

2 What is fresh water and where can it be found?

3 What are the **two** main processes in the UK used to make potable water from fresh water?

4 How are insoluble solids removed from fresh water to make potable water in the UK?

5 Suggest an alternative to adding chlorine to UK potable water.

Desalination

6 When is desalination used to make potable water?

7 Why is desalination an expensive way to make potable water?

8 What **two** processes can be used to desalinate water to make potable water?

10 Waste water

Sources of waste water

Urban lifestyles, as well as **industrial processes**, produce a large volume of **waste water**. Waste water is water that has been used in the home, business or as part of an industrial or agricultural process. Waste water must be treated before being released into the **environment** in order to prevent **pollution**.

The **urban water cycle** shows how water resources that people and industry depend on are used.

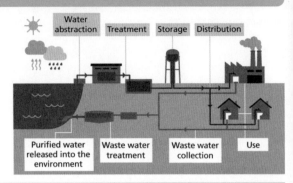

Treatment of waste water

Different sources of waste water need different **treatments**:

- **Sewage** (water that goes into sewers) and agricultural waste water require removal of **organic matter** and harmful **microbes**.
- Industrial waste water must have harmful chemicals removed and may also require the removal of **organic matter**.

Sewage water is used water from our homes that has been used in baths, sinks, washing machines, dishwashers and toilets. Sewage water in the UK is treated before the water is returned to rivers or the sea. The treatment includes:

- **Screening** and **grit** removal takes out large insoluble particles.
- **Sedimentation** lets the sewage **sludge** (solid) settle to the bottom and **effluent** (liquid) collect at the top.
- **Bacteria** are used to **anaerobically** digest sewage sludge.
- **Aerobic** bacteria are used to complete a biological treatment of effluent.

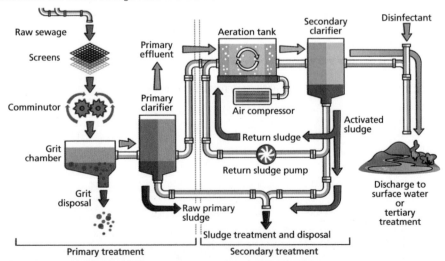

Waste water is difficult to make into potable water because several processes are needed as it contains different types of solid waste and pollutants, as well as harmful microbes.

(10) Waste water

Sources of waste water

1 Why must waste water be treated?

Treatment of waste water

2 What needs to be removed from sewage waste water?

3 What needs to be removed from waste water that comes from industry?

4 What does screening and grit removal remove from sewage?

5 **a)** What is effluent?

b) What is sludge?

c) How are effluent and sludge produced from sewage?

6 What digests sewage sludge?

7 What is used to complete the biological treatment of effluent?

8 What type of microbe is used in sewage treatment?

Alternative methods to extract metals

Ores

Metal **ores** are a **finite natural resource**. Traditional methods of extracting the metals involve:
- finding the metal ore in the Earth's surface
- mining the ore by digging, moving and disposing of large amounts of rock
- processing the ore and reduction of the metal compounds to extract the metals.

Copper (Cu) is an important metal used for electrical wires, water pipes and cooking pots. Copper ores are becoming more difficult to find and are lower-grade (have a lower percentage of copper in them), which means new ways of extracting copper from low-grade ores are needed.

Unreactive metals like gold are found as uncombined metals in nature. But most metals are found in metal compounds called minerals. When the percentage of metal in the mineral is enough to make it economical to extract the metal, we describe the mineral as an ore.

Metal solutions

Some plants can absorb metal ions as they take in water from their roots. This process can be used in **phytomining** to extract metals from low-grade ores:
- Plants absorb metal ions, e.g copper(II), Cu^{2+}.
- Plants are harvested.
- Plants are combusted.
- Remaining ash is reacted with acid to make a leachate (solution of metal compounds), e.g. react with sulfuric acid, H_2SO_4, to make copper sulfate, $CuSO_4$.

Plants absorbing metal ions

			Copper metal
Soil containing low percentage of copper ore	Plants are burnt in air	Ash containing high percentage of copper compound	

Some **bacteria** can absorb metal ions. This process can be used in **bioleaching** to make a **leachate** solution that contains metal compounds.

Both phytomining and bioleaching are time-consuming, but can be used to conserve metal ores as well as clean up contaminated soils.

Extracting the metal

The metal ions in the leachate solution must be **reduced** to make metal atoms. This can be achieved by **displacement reactions** or **electrolysis**.

Displacement reactions are when a more reactive metal takes the place of the less reactive metal in a compound. The copper containing leachate from phytomining and bioleaching can be reacted with scrap iron to displace copper from its compound:
- Word equation:
  copper(II) sulfate + iron → copper + iron(II) sulfate
- Balanced symbol equation:
  $CuSO_4 + Fe \rightarrow Cu + FeSO_4$
- Ionic equation:
  $Cu^{2+} + 2e^- \rightarrow Cu$

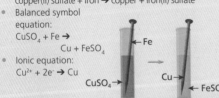

Electrolysis of the leachate can be used to extract the pure metal.

Electrolysis is also used to purify copper made from displacement as pure copper is very useful in electronics as impurities can increase electrical resistance.

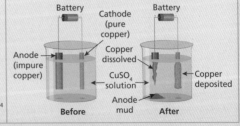

⑩ Alternative methods to extract metals

Ores

1 What type of natural resource are metal ores?

2 What are low-grade ores?

3 Give **one** use of copper metal.

Metal solutions

4 **a)** What metal extraction method uses plants?

b) What metal extraction method uses bacteria?

5 What is a leachate?

6 **a)** Give an advantage of using new ways of extracting metals from low-grade ores.

b) Give a disadvantage of using new ways of extracting metals from low-grade ores.

Extracting the metal

7 What is a metal displacement reaction?

8 **a)** What happens to the metal ions in the leachate when they undergo a displacement reaction?

b) What happens to the metal ions in the leachate when they undergo an electolysis reaction?

Life cycle assessment

The environmental impact of products and services can be assessed using a process called a **life cycle assessment** (LCA).

The LCA considers the environmental impact, including the transport and distribution, of:
- extracting and processing raw materials
- manufacturing and packaging of a product
- use and operation of a product during its lifetime
- disposal at the end of its useful life.

Scientists like to be **objective**, which means that facts are represented without bias because of personal feelings or opinions. The LCA is not purely objective because allocating numerical values to pollutant effects is less straightforward and more **subjective** as they require value **judgements**.

LCAs of different shopping bags can be used to justify which type of bag is best in terms of environmental impact.

Researchers can then use data and models to answer the investigation points considering each stage of the LCA in turn:
- Stage 1 raw materials – Paper is made from wood, which is sustainable and renewable. Plastic is made from crude oil, which is finite.
- Stage 2 manufacturing – Most energy is needed to make plastic from crude oil. Paper made from trees requires more energy than making recycled paper.
- Stage 3 use – Plastic bags are lighter and require less energy to transport compared to paper. Plastic bags are durable and can be re-used many times; paper bags are often single use.
- Stage 4 disposal – Plastic bags can always be recycled, but if they are put in landfill, they will not biodegrade. Some paper bags can be recycled up to 7 times and will biodegrade in landfill.

Researchers then must arrive at a **judgement** using the LCA. The data suggest that plastic bags have lower impact on the environment than paper bags as long as they are used many times and then recycled. The key is long life of the product and responsible disposal.

Reduce, re-use and recycle

For more **sustainable** living we should:
- **Reduce** the amount of materials we use, e.g. use thinner aluminium foil.
- **Re-use** products, e.g. a glass bottle can be cleaned and sterilised then re-used or metals can be recycled by melting and recasting or reforming into different products.
- **Recycle** materials, e.g. glass is crushed and melted; metals are melted and recast or reformed.

- Sometimes recycled materials can be added with new materials, e.g. scrap steel can be added to iron from the blast furnace, reducing the amount of new iron ore needed to make iron.

Marketing and industry can summarise LCAs to ensure they produce a favourable, pre-determined conclusion. This can lead to supported claims for advertising purposes, which are not objective or comprehensive.

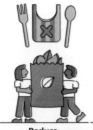

Reduce

Re-use

Recycle

10 Life cycle assessment

Life cycle assessment

1 What does an LCA measure?

2 **a)** What parts of the LCA are subjective?

b) What parts of the LCA are objective?

c) Why are some parts of the LCA objective?

3 What can happen to a paper bag at the end of its useful life?

4 What is a judgement?

Reduce, re-use and recycle

5 Describe what must be done to a glass bottle before it can be re-used.

6 How is metal recycled?

⑩ Corrosion

Corrosion

Corrosion happens when materials are destroyed by chemical reactions with the environment. When metals corrode, the metal **atoms** are **oxidised** to form metal **ions**.

Other substances like ceramics and polymers can also corrode. Corrosion causes a change in the material's properties, which may reduce its ability to complete its function. By carefully choosing materials such as **alloys**, the rate of corrosion can be reduced.

Rusting

Rusting is the name given to corrosion of iron (Fe) in the presence of water and oxygen. Rust is a brown solid made of hydrated iron(III) oxide, $Fe_2O_3.xH_2O$

Oxygen from air, and water must both be present for rusting to occur. Sodium chloride (common salt) is a catalyst for this reaction. The chemical equation for rusting is:

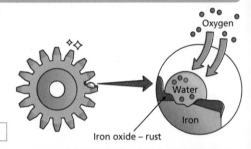

Iron oxide – rust

iron + water + oxygen ➜ hydrated iron(III) oxide

Preventing corrosion

If a barrier is applied to the material, it can prevent substances from the environment coming into contact with the material so corrosion is prevented.

Barrier prevention of rusting includes:
- adding grease onto moving parts
- painting the surface
- adding a coating of a less reactive metal using electroplating
- adding a plastic coating.

Sacrificial protection is when a more reactive metal coats the surface of the material: this more reactive metal will corrode instead of the less reactive metal.

The aluminium oxide layer can be made deeper and even coloured by using electrochemistry. This process is called **adonising**.

Aluminium (Al) is quite high in the **reactivity series** but can be used in everyday life and appears not to react. This is because aluminium is naturally covered in an oxide layer which protects the rest of the metal from corroding.

Iron and steel objects can be coated in zinc, which forms **galvanised iron**. Zinc is more reactive than iron and reacts in preference to the iron, so galvanising offers both barrier protection and sacrificial protection.

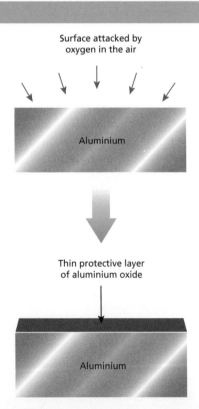

Surface attacked by oxygen in the air

Aluminium

Thin protective layer of aluminium oxide

Aluminium

10 Corrosion

Corrosion

1 What happens to metal atoms when they corrode? Tick the correct answer.

They reduce ☐

They break apart ☐

They oxidise ☐

They form a bond ☐

2 Why is corrosion a problem?

....

Rusting

3 **a)** Which element can rust?

....

b) What is the chemical name of rust?

....

c) What must be present for rusting to happen?

....

d) What is the chemical formula for rust?

....

Preventing corrosion

4 How does barrier protection prevent corrosion?

....

....

5 Suggest **one** thing that can be used as a barrier protection for metal.

....

6 How does sacrificial protection prevent corrosion?

....

....

7 Why is aluminium less reactive in real life than what you would expect from its position in the reactivity series?

....

....

....

Alloys

Pure metals are very soft. So, in everyday life, most metal materials are **alloys**.

An alloy is:
- held together by **metallic bonds**
- more durable and harder than pure metals
- a **mixture** containing mainly metals
- a **formulation**.

Common copper-containing alloys include:
- **bronze** – made from copper and tin
- **brass** – made from copper and zinc.

Aluminium is an important metal used in the aerospace industry. Pure aluminium is rarely used; instead carefully engineered aluminium alloys are used that have low density and a high strength to weight ratio.

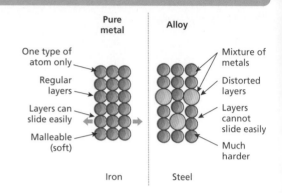

Pure metal | Alloy
- One type of atom only
- Regular layers
- Layers can slide easily
- Malleable (soft)

Iron

- Mixture of metals
- Distorted layers
- Layers cannot slide easily
- Much harder

Steel

Gold

Gold is a soft, yellow metal that can be found uncombined in nature. Gold is refined and then tested and can be sold in different amounts for making jewellery.

Gold jewellery is not made of pure gold as it would be too soft. Instead, a harder gold alloy containing silver, copper and zinc is used.

The proportion of gold in the alloy is measured in **carats**:
- 24 carat is pure gold.
- 18 carat has 75% gold.

Steel

Steel is a family of alloys made mainly of iron.

Different amounts of carbon and other metals are added to generate different properties for the different uses of each type of steel:
- High-carbon steel is strong but brittle.
- Low-carbon steel is softer and more easily shaped.
- Stainless steels contain chromium and nickel and are hard and resistant to **corrosion**.

Iron

+

Carbon

Steel

Alloys

1 Why is an alloy not pure?

2 Which elements is bronze made from? Tick the correct answer(s).

Iron ☐   Tin ☐

Copper ☐   Gold ☐

Zinc ☐

3 Which element(s) is brass made from? Tick the correct answer(s).

Iron ☐   Tin ☐

Copper ☐   Gold ☐

Zinc ☐

Gold

4 Why is gold jewellery not made from pure gold?

5 What percentage of gold is found in 18 carat gold?

6 Write the metals that are found in gold alloys.

Steel

7 What is the main desirable property of high-carbon steel?

8 Which type of steel is soft and easily shaped?

9 Which type of steel is resistant to corrosion?

10 What elements are found in stainless steel?

10 Ceramics, polymers and composites

Ceramics

Ceramic is a material that is not organic and not a metal. Clay ceramics are used to make pottery and bricks. These types of ceramics are made by shaping wet clay and then heating in a furnace to harden.

Glass is a family of ceramics made from heating sand with other substances to change the properties. Two types of glass are:

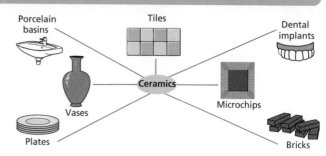

Porcelain basins

Tiles

Dental implants

Vases

Ceramics

Microchips

Plates

Bricks

- **Soda-lime glass** – made from a mixture of sand, sodium carbonate and limestone. It has a low melting temperature and is used for windowpanes and glass bottles.
- **Borosilicate glass** – often known by its brand name Pyrex. It is made from sand and boron trioxide and melts at higher temperatures. It is used in cookware.

Polymers

Polymers are long-chain molecules made from smaller repeating units called monomers.

Poly(ethene) is a polymer made from many ethene molecules joining together in an addition polymerisation reaction. The density of the packing of the polymer chains can be altered by changing the conditions of polymerisation to make two very different materials:

- **Low-density poly(ethene)** (LDPE) is made with moderate heat and high pressure. It is semi-ridged and is commonly used in single-use shopping bags.
- **High-density poly(ethene)** (HDPE) poly(ethene) is made in very high temperatures and is a tough material. It is commonly used in plastic bottles.

Thermosoftening polymers melt when heated. They have weak **intermolecular forces of attraction** between the polymer molecules, which are easily overcome when heated. This allows thermosoftening polymers to be easily reshaped and recycled.

Thermosetting polymers do not melt when they are heated. They have strong bonds between the polymer molecules, which require a lot of energy to break. This usually causes the polymer to char and combust before it would melt.

Thermosoftening polymer

The tangled web of polymer chains is relatively easy to separate.

Thermosetting polymer

Chains fixed together by strong covalent bonds – this is called **cross linking**.

The properties of polymers depend on the structure and bonding of the polymer molecule. This is directly affected by the monomers used to make the polymer (bonding) and the conditions under which they are made (structure).

Composites

Composites are:
- made of two or more distinct materials
- stronger than each material on its own.

Usually, **composites** are made from a **matrix** or **binder**, which acts like a glue sticking the other material **reinforcement** together.

Common composites include:
- reinforced concrete: matrix = concrete, reinforcement = steel
- fibre glass: matrix = polymer resin, reinforcement = glass fibre
- chipboard: matrix = resin glue, reinforcement = wood chips.

Ceramics, polymers and composites

Ceramics

1 How are clay ceramics made?

....

....

2 What are the raw materials needed to make soda-lime glass?

....

3 Why can borosilicate glass be used for cookware?

....

Polymers

4 Define the term polymer.

....

5 Why are thermosoftening polymers easy to recycle?

....

6 a) Complete the sentences by filling in the gaps.

Strong chemical bonds are found between the polymer chains in

.... polymers.

Weak intermolecular forces of attraction are found between the polymer chains in

.... polymers.

b) Label the diagrams with the correct type of polymer.

....

Composites

7 What are the **two** parts of a composite?

....

8 What is the matrix in fibre glass?

....

9 What is desirable about a composite compared to each individual material?

....

....

10 The Haber Process

The Haber Process

Ammonia (NH_3) is an important chemical used to make nitrogen-based **fertilisers**. Ammonia is made in the **Haber Process**, which has several stages:

- Air is purified to make the **reactant** nitrogen.
- Methane and steam undergo a **chemical reaction** to make the second reactant, hydrogen. The chemical equation is:
  methane + steam → carbon dioxide + hydrogen
  ($CH_4 + H_2O \rightarrow CO + 3H_2$).
- The reactants are heated to about 450°C, 200 atmosphere pressure and passed over an iron-based **catalyst** in the reactor. This causes some of the gases to react and make ammonia. The chemical equation is:
  nitrogen + hydrogen $\rightleftharpoons$ ammonia
  ($N_2 + 3H_2 \rightleftharpoons 2NH_3$).
- The ammonia is cooled, liquefied and removed.
- Unreacted gases are recycled into the reactor.

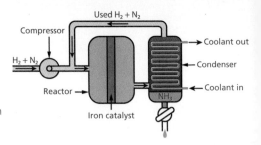

Ammonia from the Haber Process can be oxidised to form nitrogen dioxide, which reacts with water to make nitric acid. Nitric acid can react with ammonia to make ammonium nitrate, NH_4NO_3, which has a high percentage mass of nitrogen in it and would be useful in fertiliser but is also an explosive.

Fertilisers

Fertilisers are concentrated nutrients that improve plant growth and yields. NPK fertilisers contain compounds with nitrogen, phosphorus and potassium, which improve agricultural productivity.

Industrial production of NPK fertilisers uses a variety of **raw materials**:

- ammonia is reacted with acid to make ammonium **salts**
- potassium chloride and potassium sulfate can be mined
- the different chemicals are then mixed to form a specific formulation of an NPK fertiliser
- phosphate rock is **mined** and treated with acid to make **soluble** salts.

Acid	Phosphate salt
Nitric acid	Calcium nitrate and phosphoric acid (neutralised with ammonia to make ammonium phosphate)
Sulfuric acid	Single superphosphate (a mixture of calcium sulfate and calcium phosphate)
Phosphoric acid	Triple superphosphate (calcium phosphate)

In the lab, **titration** is used to add the exact amounts of base (ammonia solution) with acid (sulfuric acid). No **indicator** is used, and the solution in the conical flask should contain the salt solution that is the fertiliser. The salt can be separated from solution using **crystallisation**.

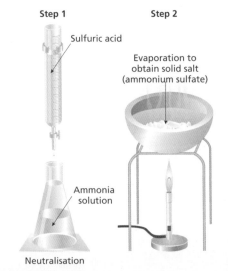

(10) The Haber Process

The Haber Process

1 What chemical does the Haber Process make?

....

2 What are the raw materials used in the Haber Process? Tick the correct answers.

Ammonia	☐	Methane	☐
Air	☐	Hydrogen	☐
Salt	☐	Fertiliser	☐
Water	☐	Acid	☐

3 What are the reactants in the Haber Process?

....

4 What type of catalyst is used in the Haber Process?

....

5 Describe the conditions for the Haber Process.

....

6 Write a balanced symbol equation for the reaction that happens in the reactor of the Haber Process.

....

Fertilisers

7 What are the elements in NPK fertilisers?

....

....

8 What **two** processes are used to make fertiliser in the lab?

....

9 In the industrial manufacture of NPK fertiliser, how are the following salts obtained?

a) Potassium

....

b) Phosphate

....

Mixed questions (paper 1)

1 Chlorine is a Group 7 element.

What is the name given to the elements in Group 7? Tick one box.

Halogens ☐

Transition metals ☐

Noble gases ☐

Alkali metals ☐

2 Chlorine has an atomic number of 17.

Draw the electronic structure of a chlorine atom.

3 **a)** Chlorine can react with hydrogen to form hydrogen chloride.

Name the bonding present in hydrogen chloride.

....

b) Give the electronic structure of a chloride ion.

....

c) Sodium can react with chlorine to form sodium chloride. Describe the bonding in sodium chloride.

....

....

4 Explain why the melting point of hydrogen chloride is lower than the melting point of sodium chloride.

....

....

....

....

....

....

Mixed questions (paper 1)

5 The Haber Process is an important industrial process to make ammonia.

Draw a dot and cross diagram of ammonia. Show only the outer shell electrons.

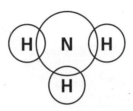

6 **a)** Ammonia can react with nitric acid to make ammonium nitrate, NH_4NO_3.

Calculate the relative formula mass of ammonium nitrate.

A_r of H = 1, N = 14, O = 16

....

b) Calculate the percentage composition of nitrogen in ammonium nitrate.

....

7 **a)** Define the term oxidation.

....

....

b) Write a word equation to show the oxidation of magnesium to form magnesium oxide.

....

8 **HT** Copper sulfate can undergo a chemical reaction with magnesium.

a) Give the name of the substance that is being oxidised.

....

b) Give the name of the substance that is being reduced.

....

c) Write a word equation for this reaction.

....

9 Combustion of natural gas is used in our homes for heating and cooking.

The figure below shows natural gas being completely combusted on a gas hob.

Draw an energy level diagram for the complete combustion of methane to form carbon dioxide and water.

Show on your diagram the activation energy and the energy given out in the reaction.

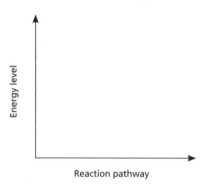

10 Compare alkanes and alkenes.

Mixed questions (paper 2)

1 **a)** Balance the symbol equation for the Haber Process.

.... N_2 + ... $H_2 \rightleftharpoons$ .... NH_3

b) Describe how ammonia is removed from the reactor in the Haber Process.

....

2 Define the term activation energy.

....

....

3 A student wanted to investigate how the concentration of hydrochloric acid affected the rate of reaction with magnesium ribbon.

The figure shows the equipment that the student used.

a) Give the independent variable for this experiment.

....

b) Give the unit of the dependent variable of this experiment.

....

c) Predict the effect of changing concentration of acid on the rate of reaction.

Use ideas from collision theory in your answer.

....

....

....

....

....

4 Crude oil is a mixture of alkanes.

The figure below is an example of an alkane found in crude oil.

```
    H   H
    |   |
H — C — C — H
    |   |
    H   H
```

a) Name the alkane in the figure above.

....

b) Give the general formula for an alkane.

....

Mixed questions (paper 2)

5 A student analysed a colourless solution to determine the ions present.

The figure below shows the bottle of unknown colourless solution.

a) The student completed a flame test on the solution. The flame turned orange-red.

Give the formula of the ion present.

_____

b) The student added some sodium hydroxide solution to a sample of the solution. Use your answer to part **a)** and suggest the observation made.

_____

_____

c) The student added acidified silver nitrate to a sample of the solution. The student observed a cream precipitate. Give the formula of the ion present.

_____

6 The atmosphere is an envelope of gas that surrounds our planet.

a) Give the name of the main element in dry air.

_____

b) Give the formula of the substance that makes up $\frac{1}{5}$ of dry air.

_____

7 For how long has the composition of the atmosphere been stable? Tick one box.

200 million years ☐

4.6 billion years ☐

2.7 million years ☐

200 billion years ☐

8 Name the atmospheric pollutant that can cause global dimming.

_____

Mixed questions (paper 2)

9 Rust is a type of corrosion that reduces the lifespan of some metal products.

The figure below shows a photograph of a steel car that has begun to rust.

a) Name the element that can undergo rusting.

_____

b) Explain why rusting is an example of an oxidation reaction.

_____

c) Car parts are often galvanised. Explain how galvanising protects a steel object from rusting.

_____

_____

_____

_____

_____

10 What process is used to separate crude oil?

_____

11 a) Alkanes are hydrocarbons. Define the term hydrocarbon.

_____

b) What process is used to make alkenes?

_____

Required practical 1

Soluble salt preparation

Aim: To prepare a pure, dry sample of a soluble salt.

REQUIRED PRACTICAL	
Preparation of a pure, dry sample of a soluble salt from an insoluble oxide or carbonate.	
Sample method	**Hazards and risks**
1. Add the metal oxide or carbonate to a warm solution of acid until no more will react. 2. Filter the excess metal oxide or carbonate to leave a solution of the salt. 3. Gently warm the salt solution so that the water evaporates and crystals of salt are formed.	• Corrosive acid can cause damage to eyes, so eye protection must be used. • Hot equipment can cause burns, so care must be taken when the salt solution is warmed.

Key points:

- A neutralisation reaction takes place between an acid and a base.
- The base (a substance that reacts with acid) is an insoluble oxide or insoluble carbonate.
- The acid is warmed to increase the rate of reaction.
- Pure crystals of the salt can be made by patting the crystals dry with absorbent paper or putting in a drying oven to remove the remainder of the solvent.

Expected results:

- Pure dry crystals of a soluble salt.

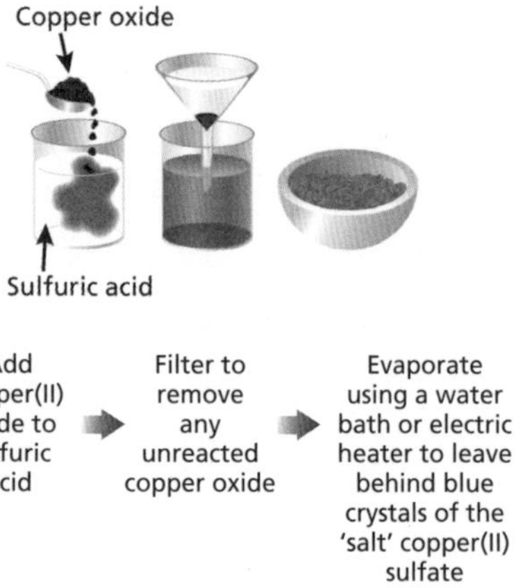

Copper oxide

Sulfuric acid

Add copper(II) oxide to sulfuric acid ➡ Filter to remove any unreacted copper oxide ➡ Evaporate using a water bath or electric heater to leave behind blue crystals of the 'salt' copper(II) sulfate

Required practical 2

Titration

Aim: To find out the reacting volumes and concentration of solutions of a strong acid and a strong alkali by titration.

REQUIRED PRACTICAL

Determination of the reacting volumes of solutions of a strong acid and a strong alkali by titration.

Sample method

1. Wash and clean dry a pipette with the alkali being used.
2. Use the pipette to measure out a known and accurate volume of the alkali.
3. Place the alkali in a clean, dry conical flask.
4. Add a suitable indicator, e.g. phenolphthalein.
5. Place the flask on a white tile so the colour can be seen clearly.
6. Fill a clean, dry burette with the acid.
7. Take a reading of the volume of acid in the burette (initial reading).
8. Carefully add the acid to the alkali, swirling the flask to thoroughly mix.
9. Continue until the indicator just changes colour. This is called the end point.
10. Take a reading of the volume of acid in the burette (final reading).
11. Calculate the volume of acid (titre) added (i.e. subtract the initial reading from the final reading).
12. Repeat the process until you have two concordant (within 0.1 cm³) titres.

Hazards and risks

- Acids and alkalis can damage the skin or eyes, so eye protection must be worn and any spillages wiped up.

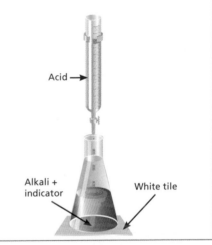

Acid →

Alkali + indicator

White tile

Key points:

- The strong acid in the burette prevents the tap of the burette from clogging up.
- A white tile is used under the conical flask to make the colour change easier to see.
- An average (mean) titre is calculated from two concordant results, that is two titre values within 0.1 cm³ of each other.
- For safety reasons, the burette should not be filled above head height and a pipette filler should be used to fill the pipette.
- For Foundation tier, a known volume of alkali is measured out using a pipette and then the burette is used to measure the volume of acid needed to fully neutralise the alkali.

HT • For Higher tier, students calculate the concentration of the alkali from the mean titre, concentration of the alkali and volume of the alkali used.

Expected results:

	Rough titration	Accurate titration 1	Accurate titration 2
Final burette reading / cm³	37.60	36.20	38.40
Initial burette reading / cm³	1.80	0.00	2.10
Volume of acid used / cm³	35.80	36.20	36.30

Break down the calculation:

1. Write down a balanced equation for the reaction to determine the ratio of moles of acid to alkali involved.
2. Calculate the number of moles in the solution of known volume and concentration. You can work out the number of moles in the other solution from the balanced equation.
3. Calculate the concentration of the other solution.

Required practical 3

Electrolysis

Aim: To investigate the electrolysis of different aqueous solutions with inert electrodes.

REQUIRED PRACTICAL	
Investigate what happens when aqueous solutions are electrolysed using inert electrodes.	
Sample method	**Hazards and risks**
1. Set up the equipment as shown in the diagram. 2. Pass an electric current through the aqueous solution. 3. Observe the products formed at each inert electrode.	• A low voltage must be used to prevent an electric shock. • The room must be well ventilated, and the experiment must only be carried out for a short period of time, to prevent exposure to dangerous levels of gas.

Key points:

- Inert electrodes do not take part in the chemical reaction. They can be made from unreactive metals like platinum but are more usually made from carbon graphite as it is cheap.
- Electrolytes must be ionic compounds where the ions are free to move and carry the charge.
- At the anode (positive electrode):
  - halogens are oxidised if the electrolyte is a halide
  - oxygen is made if the electrolyte does not contain a halide.
- At the cathode (negative electrode):
  - metals below hydrogen in the reactivity series are reduced
  - hydrogen is released if the metal is above hydrogen in the reactivity series.

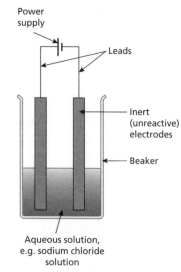

Power supply

Leads

Inert (unreactive) electrodes

Beaker

Aqueous solution, e.g. sodium chloride solution

Expected results:

- Effervescence will be observed at the electrodes if gases are made. These gases can be collected under displacement and tested:
  - hydrogen: lighted splint causes a pop
  - oxygen: relights a glowing splint
  - chlorine: damp litmus paper is bleached.
- Colour change on the cathode as solid metal is deposited if the metal in the electrolyte is below hydrogen in the reactivity series.

Required practical 4

Energy transfer

Aim: To determine the energy change that occurs in a solution-based chemical reaction.

REQUIRED PRACTICAL	
Investigate the variables that effect temperature changes in reacting solutions.	
Sample method	**Considerations, mistakes and errors**

1. Set up the equipment as shown.
2. Take the temperature of the liquid.
3. Add the other reactant, e.g. metal powder and stir.
4. Record the highest temperature that the reaction mixture reaches.
5. Calculate the temperature change for the reaction.
	• For metal and acid reactions, there should be a correlation between the reactivity of the metal and temperature change, i.e. the more reactive the metal, the greater the temperature change.
	• When a measurement is made there is always some uncertainty about the results obtained. For example, if the experiment is repeated three times and temperature changes of 3°C, 4°C and 5°C are recorded:
	  – the range of results is from 3°C to 5°C
	  – the mean (average) = $\frac{(3 + 4 + 5)}{3}$ = 4°C.

Variables	**Hazards and risks**
In the investigation between the reaction of metals and acids: • The independent variable is the metal used. • The dependent variable is the temperature change. • The control variables are the type, concentration and volume of acid.	• There is a low risk of a corrosive acid damaging the experimenter's eye, so eye protection must be used.

Key points:

- The biggest source of error is heat lost to the surroundings.
- The reaction mixture should be stirred to ensure that the temperature is the same throughout the mixture.
- A Pyrex beaker is added to stabilise the equipment and reduce the likelihood that it will fall over.

Expected results:

- A temperature rise indicates that the reaction is exothermic.
- A temperature fall indicates that the reaction is endothermic.
- Exothermic reactions include neutralisation reactions between:
  - acids reacting with metals
  - acids reacting with insoluble bases like carbonates
  - acids reacting with soluble bases or alkalis.
- Displacement of metals also causes a rise in temperature, indicating that the reaction is exothermic.

Add the metal powder

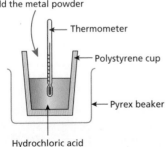

Thermometer

Polystyrene cup

Pyrex beaker

Hydrochloric acid

Required practical 5

Rate of reaction

Aim: To investigate how varying the concentration of a solution affect the rates of reactions.

REQUIRED PRACTICAL	
Investigate how changes in concentration affect the rates of reactions by methods involving the production of gas or a colour change.	
This investigation uses the reaction between sodium thiosulfate and hydrochloric acid. **Sample method** 1. Set up the equipment as shown. 2. Add the hydrochloric acid to the flask and swirl to mix the reactants. 3. Start the timer. 4. Watch the cross through the flask. 5. When the cross is no longer visible, stop the timer. 6. Repeat the experiment using hydrochloric acid of a different concentration.	**Considerations, mistakes and errors** • There should be a correlation between the concentration of the acid and the time taken for the cross to 'disappear'. • The higher the concentration of the acid, the faster the rate of reaction, and the shorter the time for the cross to 'disappear'.
Variables • The independent variable is the concentration of the acid. • The dependent variable is the time it takes for the cross to 'disappear'. • The control variables are the volume of acid and the concentration and volume of sodium thiosulfate.	**Hazards and risks** • Corrosive acid can damage eyes, so eye protection must be used. • Sulfur dioxide gas can trigger an asthma attack, so the temperature must always be kept below 50°C.

Key points:

- Rate of reaction can be monitored by the change in a reactant or product.
- If a gas is involved in the reaction, a top pan balance can monitor the mass change in an open system.
- If a gas is made in a reaction, it can be collected by displacement or by a gas syringe and the volume measured.
- If a chemical reaction causes a precipitate to be formed, then turbidity (cloudiness) can be used to monitor the reaction.

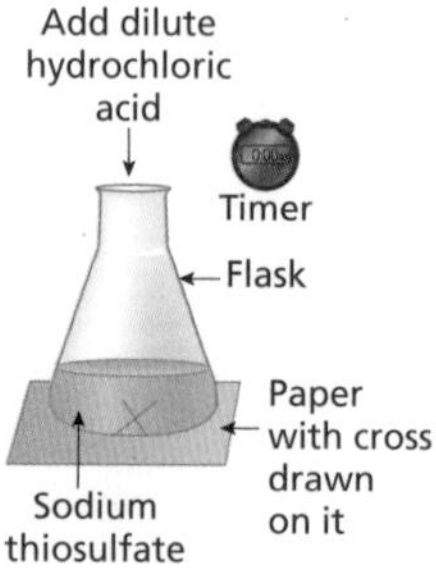

Add dilute hydrochloric acid
Timer
Flask
Paper with cross drawn on it
Sodium thiosulfate

$$Gradient = \frac{\text{difference in the amount of product formed or reactant used}}{\text{time}}$$

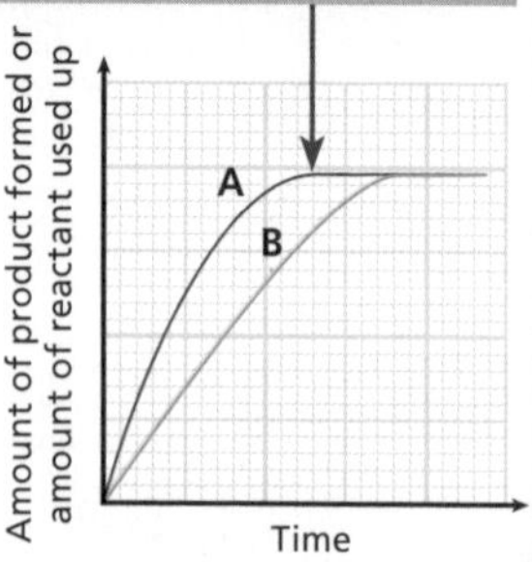

The graph shows that reaction A is faster than reaction B.

Expected results:

- The faster the mass changes or gas is collected, the faster the rate of reaction. Continuous data can be collected and a graph drawn. The steeper the gradient, the faster the rate of reaction. So, A would have a higher concentration than B, as shown in the graph.

Required practical 6

Paper chromatography

Aim: To use paper chromatography to generate and interpret a chromatogram.

REQUIRED PRACTICAL	
Investigating how paper chromatography can be used to separate and tell the difference between coloured substances.	
Sample method	**Considerations, mistakes and errors**
1. Draw a 'start line', in pencil, on a piece of absorbent paper. 2. Put samples of five known food colourings (A, B, C, D and E), and the unknown substance (X), on the 'start line'. 3. Dip the paper into a solvent. 4. Wait for the solvent to travel to the top of the paper. 5. Identify substance X by comparing the horizontal spots with the results of A, B, C, D and E.	• Only ever use pencil to draw the start line, as ink will disolve and affect your results.

Key points:

- The solvent front must be marked before the chromatogram is dried as it will not remain visible.
- A lid can be placed on the chromatography tank to slow down the wicking of the solvent and improve the separation.
- If a dye does not move from the start line, it is not soluble in the solvent.
- A pure substance will have only one dot on a chromatogram.
- Dots in different lines at the same height will be the same substance.

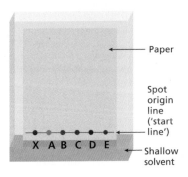

Paper

Spot origin line ('start line')

Shallow solvent

Expected results:

- A dry chromatogram which shows the separation.
- The R_f values of each spot can be calculated. For example, the blue spot R_f value = $\frac{4}{10}$ = 0.4.

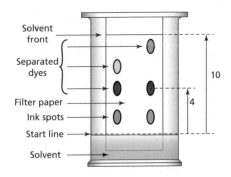

Solvent front

Separated dyes

Filter paper

Ink spots

Start line

Solvent

Required practical 7

Chemical analysis

Aim: To use chemical tests to identify the ions in unknown single ionic compounds.

REQUIRED PRACTICAL	
Identify the ions in a single ionic compound using chemical tests, e.g. flame tests.	
Sample method	**Hazards and risks**
1. Heat a piece of inert wire (e.g. nichrome or platinum) in a Bunsen flame and then dip it in concentrated hydrochloric acid to clean it. 2. Dip the wire in the compound. 3. Put it into a blue Bunsen flame and observe what colour flame is produced.	• There is a risk of being burned by the hot wire so care must be taken not to touch it. • The concentrated acid is corrosive so avoid skin contact and use eye protection. • The compounds used may be harmful so avoid skin contact.

Key points:

- If the flame test did not give a positive result, add sodium hydroxide solution to a fresh sample. If there is a white precipitate then calcium or magnesium ions could be present, a blue precipitate is copper(II) ions, green precipitate is iron(II) and a brown precipitate is iron(III).
- In fresh samples of solution add:
  - acidified silver nitrate and observe: if there is a white precipitate then chloride ions are present; if there is a cream precipitate then bromide ions are present; if there is a yellow precipitate then iodide ions are present
  - acidified barium chloride: if there is a white precipitate then sulfate ions are present
  - a dilute acid and if effervescence is observed, test the gas with limewater. If the limewater turns cloudy then carbonate ions are present.

Expected results:

- A coloured flame or a coloured precipitate with sodium hydroxide to identify the metal ion present in the ionic compound.
- A precipitate with either acidified silver nitrate or acidified barium chloride or effervescence with a dilute acid to determine the non-metal ion that is present.

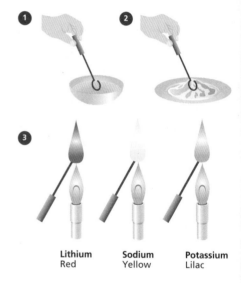

Lithium	**Sodium**	**Potassium**
Red	Yellow	Lilac

Required practical 8

Water

Aim: To analyse and purify water samples.

REQUIRED PRACTICAL	
Analysis and purification of water samples from different sources.	
Sample method	**Hazards and risk**
1. Use a pH probe or suitable indicator to determine the pH of the sample. 2. Set up the equipment as shown. 3. Heat a set volume of water sample to 100°C so that the water changes from liquid to gas. 4. The pure water collects in the condenser and changes state from gas to liquid. Collect this pure water in a beaker. 5. When all the water from the sample has evaporated, measure the mass of solid that remains to find the amount of dissolved solids present in the sample.	• There is a risk of the experimenter burning themselves on hot equipment, so care must be taken during and after the heating process.

Key points:

• pH can be measured using universal indicator and comparing the colour that it turns with the indicator chart.
• pH can also be measured with a pH probe. If the pH probe is connected to a machine to record the data, then it becomes a datalogger.
• The mass of the dissolved solids is measured using a top pan balance.

Expected results:

Water sample	pH	Mass of solid obtained by evaporating 50 cm³ of water sample (g)
Distilled	7	0.00
Mineral	9	1.74
Carbonated	4	0.07

• Different samples of water have different masses of residue in them and only pure water will have no residue. As water samples are usually mixtures, their pH varies between different sources.

Working scientifically

Chemistry

Chemistry is a type of science and the methodical study of the structure and behaviour of matter. This involves suggesting hypotheses to explain facts and observations. A hypothesis must be able to be tested by experiments.

Sometimes chemists design experiments to test hypotheses and theories by using:
- their own senses through observation
- measuring instruments to make observations.

Scientific method

Science is the study of the physical and natural world through **observation** and **experiment**. The **scientific method** is the way that scientists work in a **systematic approach**.

At the start of an investigation, you should set an **aim**. This is what you want to find out in your investigation and is often a question. All of the **data** that you collect in your investigation should help you answer the aim and is called **valid** data.

> There is no need to collect data that doesn't help you answer the aim. For example, if you wanted to know the average height of your classmates you need to collect quantitative data of their heights, but you do not need to collect data on their hair colour as hair colour is not valid data that would help you answer your aim.

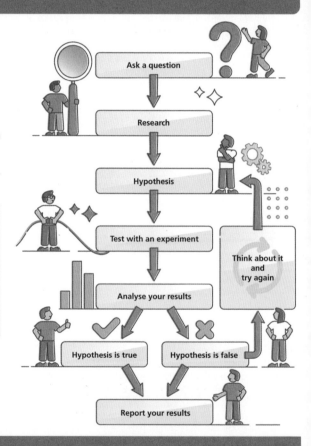

- Ask a question
- Research
- Hypothesis
- Test with an experiment
- Analyse your results
- Hypothesis is true
- Hypothesis is false
- Think about it and try again
- Report your results

Theories

When an explanation applies to lots of different situations and examples, it is described as a theory. Scientific theories, such as how the atmosphere developed, have changed over time. It is very difficult to collect evidence from events that happened so long ago. In addition, technology is developing all the time, which allows new data to be collected and analysed.

Data can be interpreted in different ways and some data is more reliable and accurate than other data. So, it is important that data and conclusions are peer reviewed (checked by other scientists) before being shared with the public. This prevents bias that could lead to misleading conclusions that could cause harm.

Working scientifically

Types of research

Chemists ask questions. Using the knowledge they already have, chemists have ideas about what the answers to the questions might be. To test these ideas, chemists carry out **scientific investigations**, in which they collect **data**.

There are two main types of investigation:
- **primary research** – where data is collected first hand
- **secondary research** – where data is used that other people have collected (like those found in books, trusted internet sites and data your teacher has collected).

For primary research you could complete:
- **an experiment**, where you change the independent variable and collect data on the change observed in the dependent variable
- **a survey**, where you monitor something that is already happening.

The research tries to find the relationship between the:
- **independent variable** – the variable you choose to change
- **dependent variable** – the variable that you measure during the investigation.

To make this a **fair test** and give **valid results**, some **variables** must be identified and all efforts made to keep them constant in each run of the experiment. These variables are called **control variables**.

Risk assessment

Hazards are the damaging effects that are potentially possible from the chemicals and procedures used in an investigation. The **risk** is the likelihood of the damaging effects happening during the investigation. So, a **risk assessment** identifies the hazards, classifies the likelihood of any issues arising and makes suggestions to reduce the potential harm caused. For example, getting acid in the eye could cause damage; by wearing PPE and using small volumes of acid, the risk is mitigated. But if the risk is too high, it may not be possible to safely complete the investigation at all.

Measuring equipment

When you are choosing to measure quantities it is important to choose the correct measuring equipment that:
- has the **correct range** – it can measure all the values that you need for your investigation
- is **accurate** – the value measured is close to the true value
- has suitable **interval** and **resolution** – it can detect the changes and you can read the values.

> Choose a measuring instrument that has the biggest interval and the smallest range for what you are trying to measure, so that you get the most accurate measurement.

Quantity to be measured	Measuring equipment
Volume – measure the volume at the bottom of the meniscus and in your eye line.	Measuring cylinder, burette, pipette
Temperature – make sure the bulb of the thermometer is in the substance you want to take the temperature of.	Thermometer
Time – remember to choose seconds or minutes to record your time. (1:30 on a stopwatch is 1 minute and 30 seconds, which is 90 seconds or 1.5 minutes.)	Stopwatch
Mass – in a lab, a top pan balance is used.	Top pan balance

Working scientifically

Method

The **method** is a step-by-step description of the investigation. It should be written in such a way that anyone could read it and complete the same investigation. If the results are similar each time the method is repeated, they are described as **reliable** results.

Recording data

During an experiment, data is collected and recorded into a results table. This should be drawn in advance so that only the dependent variable results are written down during the investigation.

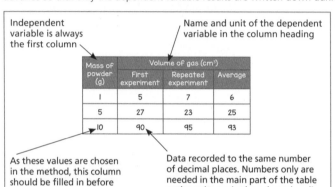

Independent variable is always the first column

Name and unit of the dependent variable in the column heading

Mass of powder (g)	Volume of gas (cm³)		
	First experiment	Repeated experiment	Average
1	5	7	6
5	27	23	25
10	90	95	93

As these values are chosen in the method, this column should be filled in before the investigation starts

Data recorded to the same number of decimal places. Numbers only are needed in the main part of the table as the units are in the column heading

Researchers must keep all the other control variables the same to make the data valid. Valid data is information that can be used to find out if the scientist's idea that is being investigated is correct or incorrect.

In a scientific survey, it is often useful to collect the data in a **frequency table** (**tally chart**).

Types of data

- **Qualitative** – using your senses to make observations; this usually includes what you see, what you hear and what you smell.
- **Quantitative** – numbers and values that you have read and recorded from measuring instruments.

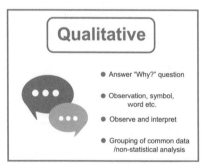

Qualitative
- Answer "Why?" question
- Observation, symbol, word etc.
- Observe and interpret
- Grouping of common data /non-statistical analysis

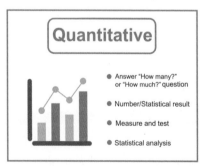

Quantitative
- Answer "How many?" or "How much?" question
- Number/Statistical result
- Measure and test
- Statistical analysis

Scientists prefer to collect quantitative data because it is:
- **repeatable** and **reproducible** – each time the investigation is repeated, similar results are collected
- **objective** – the data is not influenced by how the person is feeling
- **useful** – graphs and charts can be made which allow **predictions** outside the data collected.

Working scientifically

Good data

Scientists want to collect **accurate** data that is close to the true value. But they also want their investigations to produce **reliable** data (data that is similar each time the experiment is repeated).

Imagine if every time you threw a dart at a dartboard you got the bullseye. You would be both reliable and accurate in your throwing.

But if sometimes you hit the bullseye and sometimes you don't, then you are sometimes accurate but you are not reliable.

Precise data have a small range (very little spread) and are similar to the **mean** value. **Random errors** affect precision, but it gives no indication of how close the results are to the true value.

How accurate are my results?

An **accurate result** is close to the accepted value. The accepted value could be:
- written in a text book
- found on a trusted website
- the results collected by your teacher.

The accuracy of the results depends on:
- the quality of the **measuring instruments**
- the skill of the researcher in completing the practical.

A researcher can determine how accurate their results and conclusions are by comparing them to the accepted value.

Maximum accuracy

Accuracy of results can be improved by:
- carefully following the **method** for the investigation to reduce **random errors**
- correcting any data for systematic errors – remember to calibrate all the measuring equipment
- using an average of the data – take repeated measurements, remove any anomalous results and calculate the **mean**
- selecting a measuring instrument with a **scale** that covers the **range** of values needed for the experiment and is also able to detect the smallest possible change in a measurement.

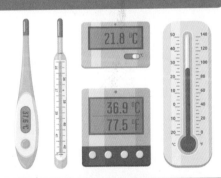

Reliable (similar each time you repeat) and accurate (close to the true value) results are the goal of a researcher.

Maths skills

Units

Units are used so that measurements can be compared. There are **quantities** that are often measured in science and you need to know their standard units:

Variable	Unit	Symbol
time	seconds	s
length	metres	m
mass	grams	g
volume	cubic centimetres	cm³

Prefixes

In science, we often work with numbers that are very big or very small. **Prefixes** can be used to make the numbers more manageable and scientists change the units so that the number is between 1 and 100.

Think about a common variable that you might measure and how you could use the prefixes to change the unit by a factor of ten.

Some examples you are likely to come across include:
- a **kilo**gram is one thousand **grams**
- a **centi**metre is one hundredth of a **metre**
- a **milli**litre is a thousandth of a **litre**.

Time

In investigations, stopwatches or timers are often used to measure time.

Consider a stopwatch that displays 1:30; we read this as 1 minute and 30 seconds, but this is actually two units. So the value needs to be converted either into 90 seconds or 1.5 minutes. (It should never be written as '1.30 minutes', which is actually 78 seconds!)

Maths skills

What is a graph?

A **graph** is a type of visual representation that looks at the numerical relationship between two numerical quantities. In a graph:
- both variables are continuous
- the independent variable is plotted on the x-axis
- the dependent variable is plotted on the y-axis.

When plotting a graph, choose scales carefully so that the data is plotted over at least half of the graph paper. This makes it easier to spot any patterns.

Plot the points with crosses (x) as this is more accurate.

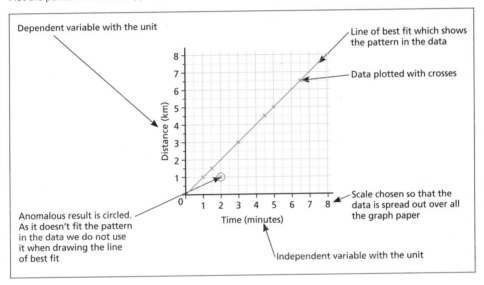

Dependent variable with the unit

Line of best fit which shows the pattern in the data

Data plotted with crosses

Scale chosen so that the data is spread out over all the graph paper

Anomalous result is circled. As it doesn't fit the pattern in the data we do not use it when drawing the line of best fit

Independent variable with the unit

A graph allows us to see a pattern in the data and to make **predictions**. Points that do not fit the pattern are called **anomalous points**. Hold your graph at arm's length and look at the points:
- Do they form a straight line? Use a ruler to draw a straight line of best fit going through as many of the points as you can.
- Do they form a curve? In a smooth arc, try to draw a curve with just one stroke of the pencil.
- Do they form any pattern at all? If there is no pattern, do not draw a line of best fit.

Make sure that you use a sharp pencil and draw one clear **line of best fit**. It is important not to sketch or make a thick line, because the predictions that you make from these lines will be less useful.

> In science, line of best fit means the line that shows the relationship between the independent variable and the dependent variable. So, in science, lines of best fit can be straight lines, curves or even shapes.

Maths skills

Statistical analysis

Quantities are expressed in numbers. It is useful to express these in digits that suggest that the **magnitude** (size) is **accurate**. Significant figures usually start with the first non-zero number of the quantity and then the number is rounded at that point to show the confidence in number.

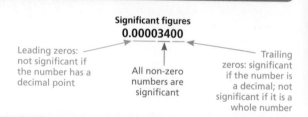

Significant figures
0.00003400

Leading zeros: not significant if the number has a decimal point

All non-zero numbers are significant

Trailing zeros: significant if the number is a decimal; not significant if it is a whole number

Range and averages

The **range** of data shows how spread out it is. A small range of data suggests that it is more consistent.

Often, three sets of data are collected for each experiment. If all the values are similar, then the data is reliable. But if a value is not similar you should disregard it as an **anomaly**. There will still be some variation in the results so it is usually worthwhile calculating the **average**.

- **Mean** – the sum of the numbers divided by the amount of numbers
- **Median** – the number in the middle
- **Mode** – the number that appears the most
- **Range** – the difference between the greatest and smallest numbers

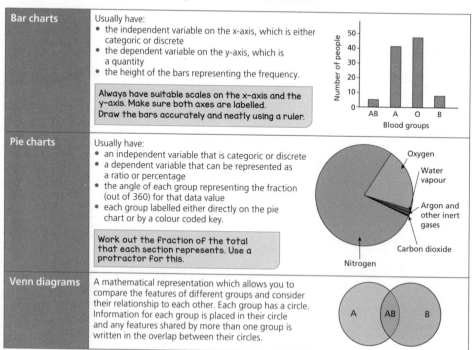

Bar charts	Usually have: • the independent variable on the x-axis, which is either categoric or discrete • the dependent variable on the y-axis, which is a quantity • the height of the bars representing the frequency. Always have suitable scales on the x-axis and the y-axis. Make sure both axes are labelled. Draw the bars accurately and neatly using a ruler.	
Pie charts	Usually have: • an independent variable that is categoric or discrete • a dependent variable that can be represented as a ratio or percentage • the angle of each group representing the fraction (out of 360) for that data value • each group labelled either directly on the pie chart or by a colour coded key. Work out the fraction of the total that each section represents. Use a protractor for this.	
Venn diagrams	A mathematical representation which allows you to compare the features of different groups and consider their relationship to each other. Each group has a circle. Information for each group is placed in their circle and any features shared by more than one group is written in the overlap between their circles.	

Answers

Page 7

1. The smallest particle of an element that can exist on its own.
2. All the atoms (particles) are the same.
3. (Just under) 100
4. neutron; proton; electron
5. A charged atom
6. **a)** Positive
   b) Negative
7. Metals
8. They are different forms of the same element.
   Or: They are atoms with the same atomic (proton) number but a different mass number.
9. **a)** They have the same number of protons in the nucleus.
      Or: They have the same number and arrangement of electrons in electron shells.
   b) They have different numbers of neutrons in the nucleus.
10. 3

Page 9

1. A chemical reaction
2. **a)** 3
   b) 2
3. Chemical reactions
4. More than one substance not chemically joined together.
5. A mixture that is designed to be a useful product.
6. Tin (Sn) and Copper (Cu)
7. **a)** Insoluble solid from a liquid / solution
   b) Distillation
   c) Similarity: Both techniques separate solutions.
      Difference: Distillation collects the solvent whereas crystallisation collects the solute.
   d) Fractional distillation

Page 11

1. To understand observations and make predictions.
2. New evidence (data) is collected.
3. Solid spheres that could not be divided.
4. Plum pudding model

5. Most of an atom is empty space.
6. A small positive nucleus is at the centre of the atom.
7. James Chadwick
8. nucleus; radius; protons/neutrons; neutrons/protons; positive/+1; neutral/0; 1

Page 13

1. **a)** Atomic mass
   b) Atomic (proton) number
2. **a)** Sodium
   b) 11
   c) 11
   d) 12
3. Number of protons and number of electrons
4. Mass number – atomic (proton) number
5. Atomic (proton) number and so the number of protons and electrons
6. The periodic table shows relative atomic mass, not atomic mass.
7. The number of protons in an atom of an element.

Page 15

1. **a)** 2   **b)** 8   **c)** 8
2. From the centre and each energy level is completed before the next one is started.
3. 1
4. 3
5.

2,1

6. Argon
7. A charged atom
8. **a)** 1+
   b)

Sodium ion
2,8

9. All ions have a complete outer shell of electrons. A sodium atom loses one electron and a fluorine atom gains one electron, resulting in both ions having the same electronic structure.

Page 17

1. Putting objects into groups with similar properties.
2. To describe objects without confusion, to see connections between different objects and to make predictions about new objects that are found and fit into a particular group.
3. Not all elements had been discovered and some elements were put in inappropriate groups.
4. **a)** Broadly by increasing atomic weight and by moving some elements around to group them by properties.
   b) He believed that more elements were still to be discovered.
   c) The discovery of elements that matched his predictions, and the discovery of subatomic particles and isotopes.
5. By increasing atomic (proton) number
6. Group 0 (noble gases)
7. solid (Accept metals)

Page 19

1. In the first column
2. Alkali metals
3. They all have one electron in the outer shell.
4. 1+
5. **a)** It decreases
   b) There is a decrease in the force of attraction between the atoms.
   c) It decreases
   d) It increases
6. Rubidium oxide (Rb_2O)
7. Burns quickly with a bright yellow flame and a white solid is produced
8. Lilac
9. An observation where you hear fizzing and see bubbles.

Page 21

1. 7th main column of the periodic table / 17th column of the periodic table
2. Halogens
3. All have seven electrons in the outer shell.
4. 1–
5. **a)** It increases.
   b) The molecules get larger and and there is a stronger attraction between them.
   c) They get darker.
   d) It decreases.

6. A 1- ion resulting from a halogen/a halogen atom gaining one electron in the outer shell.
7. A metal compound that contains a halide ion.
8. A reaction in which a more reactive element/halogen takes the place of a less reactive halogen/element from its compound.
9. chlorine + sodium bromide → sodium chloride + bromine

Page 23

1. Last column
2. Noble gases
3. All have a full outer shell of electrons / stable electron configuration.
4. Gases
5. **a)** Increases
   b) Increases
6. As relative atomic mass increases, so does the boiling point.
7. **a)** 2
   b) 8
8. They have a full outer shell of electrons / they are electronically stable
9. **a)** Do not easily form molecules
   b) Do not easily form ions

Page 25

1. Between Groups 2 and 3/in the centre of the periodic table
2. Any one of: chromium (Cr), manganese (Mn), iron (Fe), cobalt (Co), nickel (Ni), copper (Cu)
3. Any one of: conductor, ductile, malleable, lustrous (shiny)
4. **a)** Transition metals are denser.
   b) Transition metals' melting point is higher than Group 1/Group 1 metals have a lower melting point than transition metals.
   c) Transition metals are stronger and harder than Group 1 metals/Group 1 metals are weaker and softer than transition metals.
   d) Transition metals are less reactive than Group 1 metals/Group 1 metals are more reactive than transition metals.
   e) Transition metal compounds are coloured but Group 1 compounds are white.
5. **a)** copper(I) and copper(II)
   b) Cu^+ and Cu^{2+}
6. To speed up the reaction
7. manganese(VI) oxide/manganese dioxide

Page 27

1. Non-metals
2. A shared pair of electrons
3. Noble gases / Group 0
4.
5. 3
6. H–N–H
   |
   H
7. A giant structure made of very large molecules, made of smaller repeating units.
8. The atoms are too far apart from each other, and the electrons are not visualised.
9. Dot and cross diagram

Page 29

1. They lose outer shell electrons.
2. 2+
3. Na^+
4. They gain electrons into their outer shell.
5. 1^-
6. O^{2-}
7. **a)** The electrostatic force of attraction between oppositely charged ions
   b) A metal and a non-metal
   c) From the metal to the non-metal
8. A giant structure / lattice
9. Na_2O

Page 31

1. Substances that contain atoms of only one metallic element.
2. **a)** metallic
   b) giant structure
   c) in a regular pattern
3. The sharing of the delocalised electrons between the metallic ions in a giant structure.
4. Conductor of heat – Delocalised electrons can transfer the energy.
   Malleable (bends and shapes easily) – The layers of atoms easily slide over each other.
   High melting and boiling points – Many strong metallic bonds must be broken.
   Conductor of electricity – Delocalised electrons are free to move and carry charge.
5. A mixture / formulation of a metal and at least one other element.
6. Steel

Page 33

1. Shared pairs of electrons / covalent bonds
2. Lattices
3. Pure metal elements and alloys
4. Liquid and Aqueous solution
5. All of the atoms in these structures are linked to other atoms by strong bonds. A lot of energy is needed to break any bond.
6. Non-metal elements
7. The greater the size, the greater the melting and boiling points.
8. Electrical insulator – No charged particles are free to move and carry a charge.
   Low melting and boiling points – Only relatively weak intermolecular forces of attraction need to be overcome and no bonds are broken.
   Soft and brittle – Weak intermolecular forces of attraction are easily broken.

Page 35

1. liquid; gas; solid *OR* gas; liquid; solid
2. Particles are fixed and are not able to flow past each other.
3. Liquids and solids
4. To explain observations about matter and make predictions about how matter will behave if we change the conditions.
5. Buckminsterfullerene
6. No new substance is made.
7. Melting
8. A substance changes from a gas to a liquid state.
9. Gas
10. Dissolved in water / aqueous solution / solution where water is a solvent.
11. (l)

Page 37

1. 4
2. **a)** Covalent
   b) Covalent
3. **a)** 4
   b) Many strong covalent bonds must be broken to pull apart the atoms from the giant covalent structure.
4. **a)** The outer shell of each carbon atom.
   b) 3
   c) The layers / planes of atoms slide easily over each other.

5. a) One layer / plane of hexagonal rings of carbon atoms / graphite.

  b) Delocalised electrons are free to move and carry the charge.

6. C_{60}

7. Very high length to diameter ratios (*Accept* high melting and boiling points).

Page 39

1.

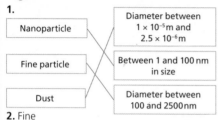

2. Fine

3. A few hundred

4. High

5. $8\,nm^3$

6. Volume ratio increases by a factor of 10.

7. *Any one of:* medicine; electronics; cosmetics; sun creams; deodorants; catalysts.

8. Very small particles can penetrate into the human body, so medicines and sun creams are more easily absorbed.

9. Possible risk of polluting the environment where the nanoparticles could build up in organisms and we are unaware of the long-term effects.

Page 41

1. On the periodic table

2. 12

3. The sum of the atomic masses for each atom in the molecule.

4. a) 2

  b) 48

5. The sum of the atomic masses for each atom in the molecule.

6 a) 34

  b) 94

7. The sum of the atomic masses for each atom in the empirical formula.

8. a) 72.7%

  b) 11%

  c) 89%

Page 43

1.

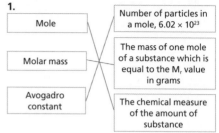

2. Number of moles = mass ÷ relative atomic mass

3. 2 moles

4. Number of moles = mass ÷ relative formula mass

5. 2 moles

6. 80 g

7. The reactant that is fully used up in the chemical reaction.

8. Add too much / so much that the reactant is still visible.

Page 45

1.

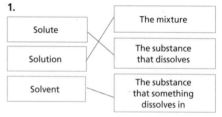

2. A lot of solute in a given volume of solvent.

3. Very little solute in a given volume of solvent.

4. Concentration = mass ÷ volume (*Accept* $C = \dfrac{m}{V}$)

5. $\dfrac{10}{2.5} = 4\,g/dm^3$

6. Concentration = amount of substance ÷ volume (*Accept* $C = \dfrac{n}{V}$)

7. $\dfrac{0.5}{2.5} = 0.2\,mol/dm^3$

8. g/dm^3 and mol/dm^3

Page 47

1. A measure of how much desirable product has been made and collected.

2. Top pan balance

3. a) %

  b) Grams, g

  c) Grams, g

4. (actual yield ÷ theoretical yield) × 100

5. 20%

6. 25 g

7. Atom utilisation

8. 100%

Page 49

1. a) 20°C
   b) 1 atmosphere / 100 kPa

2. 24 dm³ / 24,000 cm³ / 24 litres

3. × 1000

4. volume (dm³) = amount of gas (mol) × 24

5. a) 2 moles
   b) 1 mole
   c) 0.5 moles

6. Hydrogen

7. (g)

8. 0.5 moles

9. 12 dm³

Page 51

1. A new substance is made.

2. No new substance is made.

3. a) Reactants   **b)** Products

4. Goes to

5. Conservation of mass / no atoms are created or destroyed, only rearranged.

6. $2Mg + O_2 \rightarrow 2MgO$

7. The ions involved directly with the chemical reaction.

8. An ion that remains in solution and is unchanged from the reactants to the products.

9. Either oxidation or reduction.

10. $Cu \rightarrow Cu^{2+} + 2e^-$

Page 53

1. Oxygen is added to the fuel.

2. Electrons are lost.

3. Oxygen is removed from the iron oxide.

4. Electrons are gained.

5. Base and acid

6. Salt

7. Electrolysis and thermal

8. An ionic compound that is molten or in aqueous solution.

9. An insoluble solid

10. Filtering

11. A more reactive halogen will take the place of a less reactive halogen in a compound.

12. A more reactive metal will take the place of a less reactive metal in a compound.

Page 55

1. The outer shell electrons are lost.

2. $Na \rightarrow Na^+ + e^-$

3. A list of metals from most to least reactive.

4. Reaction of metals with water and reaction of metals with acids.

5. Carbon and hydrogen

6. *Any suitable example, e.g.* lead, tin, iron and zinc

7. A more reactive metal will take the place of a less reactive metal in its compound.

8. Magnesium sulfate and zinc

9. Magnesium is more reactive than zinc and magnesium is already in the compound.

10. copper + magnesium sulfate has no reaction, so no word equation.

11. magnesium + copper sulfate → copper + magnesium sulfate

Page 57

1. Uncombined

2. A compound containing a metal that is found in nature.

3. A mineral where it is economical to extract the metal.

4. Reduction

5. When the metal is below carbon in the reactivity series and the metal doesn't react with carbon.

6. iron oxide + carbon → carbon dioxide and iron

7. Global warming and climate change

8. When the metal is above carbon in the reactivity series or reacts with carbon.

9. A lot of energy is needed to melt the metal-containing compound and a lot of electricity is needed.

10. aluminium oxide → aluminium + oxygen

11. The oxygen that is made at the anode immediately reacts with the carbon of the anode and burns it away.

Page 59

1. Oxygen is lost from a substance.

2. Electrons are gained by a substance.

3. Aluminium oxide loses oxygen.

4. Hydrogen ions

5. Oxygen is gained.

6. Electrons are lost.

7. Oxygen is gained by the carbon.

8. A chemical change where both oxidation and reduction happen at the same time.

9. As one reactant is oxidised, the other reactant is reduced. So, both oxidation and reduction happen at the same time.

10. a) Magnesium

   b) copper(II) ions / copper(II) sulfate

   c) Both oxidation and reduction are happening at the same time.

Page 61

1. $H^+(aq)$

2. Sulfuric acid

3. Metals above hydrogen in the reactivity series.

4. Neutralisation

5. Calcium sulfate and water

6. Sodium chloride and water

7. potassium hydroxide + sulfuric acid → water + potassium sulfate

8. Carbon dioxide

9. sodium carbonate + nitric acid → sodium nitrate + carbon dioxide + water

10. An ionic compound where the hydrogen in an acid has been swapped out for a metal ion or ammonium ion.

11. Precipitation

Page 63

1. A measure of the acidity / alkalinity of a solution.

2. 0–14

3. Add water to dilute, or add a chemical to neutralise some of the acid or alkali.

4.

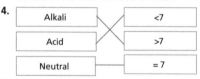

Alkali		<7
Acid		>7
Neutral		= 7

5. Universal indicator

6. pH probe

7. Acid

8. Alkali

9. Water

Page 65

1. To find the concentration of an acid or an alkali.

2. The volume of the solution in the burette at the start and end of a titration.

3. The volume of solution added from the burette.

4. From the bottom of the meniscus and in your eye line.

5. a) Close to the true value.

   b) Data reported to the same number of decimal places.

   c) Titre values within 0.1 cm³ of each other.

6. To estimate the volume of solution needed to be added from the burette.

7. The average titre calculated from accurate titrations.

8. Mean titre = sum of the two concordant results ÷ 2

9. Number of moles = concentration × volume

10. The mole ratio / the amount of acid that will react with the alkali.

Page 67

1. The process of making ions.

2. The greater the amount of $H^+(aq)$, the lower the pH value.

3. a) An acid (substance that releases $H^+(aq)$) that fully ionises in solution.

   b) *Any one of:* nitric acid; HNO_3; sulfuric acid; H_2SO_4; hydrochloric acid; HCl

4. $HA → H^+ + A^-$

5. $HCl → H^+ + Cl^-$

6. a) An acid (substance that releases $H^+(aq)$) that partially ionises in solution.

   b) *Any one of:* citric acid; ethanoic acid (CH_3COOH); carbonic acid (H_2CO_3); any organic acid

7. $HA \rightleftharpoons H^+ + A^-$

8. $CH_3COOH → H^+ + CH_3COO^-$

9. $HCOOH → H^+ + HCOO^-$

Page 69

1. a) A new substance is made.

   b) Energy is taken in by the system.

2. *Any one of:* positive ions; metal ions; hydrogen ions

3. Electrons are lost from the ions (oxidation) to form neutral atoms.

4. Direct (dc)

5. Molten (l) or aqueous solution (aq)

6. Carbon (C)

7. *Accept any metals above carbon in the reactivity series.*

8. Cathode

9. decreases; increases

Page 71

1. anode; halogen; oxygen

2. $2Br^- → Br_2 + 2e^-$

3. Pure metal

4. Hydrogen
5. Copper
6. $Cu^{2+} + 2e^- \rightarrow Cu$
7. **a)** Pb
   b) Br_2
8. Carbon in the anode reacts with the oxygen that is produced in the electrolysis and burns the anode away.
9. An ion in solution that is not directly involved in the chemical reaction.

Page 73

1. It is the same / constant.
2. Energy is released from the chemical system into the surroundings.
3. Stored chemical energy of the products is less than the reactants.
4. Thermometer
5. **a)** A diagram which models the energy changes from the start, during and at the end of the reaction.
   b) The difference between the energy level of the reactants and the products.
   c) Reactants have a higher energy level than the products.
6. Hand warmers and self-heating cans
7. H_2O
8. Aqueous solution

Page 75

1. Energy comes from the surroundings into the chemical system.
2. Stored chemical energy in the products is greater than the stored chemical energy in the reactants.
3. Temperature decreases.
4. **a)** Time / progress of the reaction
   b) Energy
5. The difference in the energy level of the reactant and the product.
6. Thermal decomposition
7. *Any one of:* sherbet in sweets; sports injury packs

Page 77

1. **a)** Chemical bonds break.
   b) Chemical bonds are made.
2. endothermic; exothermic; released
3. C=O is stronger than C–H
4. kJ/mol

5. **a)** The sum of the product bonds is stronger than the sum of the reactant bonds.
   b) The sum of the reactant bonds is stronger than the sum of the product bonds.
6. Overall energy change for a reaction = energy needed to break the reactant bonds – energy released when product bonds are made
7. **a)** exothermic reaction
   b) endothermic reaction

Page 79

1. A store of chemical energy that can be converted into electricity.
2. electrodes; metals; electrolyte
3. More than one chemical cell connected in series.
4. Alkaline battery
5. Irreversible reaction
6. They can cause pollution if they are discarded into landfill.
7. **a)** They are cheaper to produce than rechargeable cells.
   b) Non-rechargeable cells are expensive to recycle.
8. Reversible reaction
9. It can be used many times so fewer finite (non-renewable) resources are used.
10. Apply an external electrical current.

Page 81

1. To produce electricity.
2. It electrochemically oxidises.
3. Water
4. $2H_2 + O_2 \rightarrow 2H_2O$
5. **a)** Negative electrode (cathode)
   b) Positive electrode (anode)
6. **a)** $2H_2 + 4OH^- \rightarrow 4H_2O + 4e^-$
   b) $O_2 + 2H_2O + 4e^- \rightarrow 4OH^-$
7. The hydrogen fuel can be made from electrolysis of water using renewable energy.
8. They are expensive to make.

Page 83

1. The speed of the chemical reaction / the change in the reactant used or the product formed in a given time.
2. Top pan balance
3. **a)** Gas from the air is a reactant.
   b) Product is a gas and is released to the air.
4. Cloudy

5. Use the disappearing cross method.
6. g/s and cm^3/s
7. Time
8. The steeper the gradient, the faster the rate of reaction.
9. The reaction has stopped.

Page 85

1. A scientific model to explain and predict the rate of a chemical reaction.
2. A collision between reactant particles, with energy equal to or greater than activation energy and in the correct orientation.
3. increases; increases
4. There are more collisions between reactants in a given time, but the same percentage are successful, so overall there are more successful collisions in a given time and therefore a faster rate of reaction.
5. It increases.
6. Enzyme
7. They stay the same.
8. B

Page 87

1. A chemical change where the reactants make the products and the products make the reactants.
2. a) Heat b) Cooling c) 176 kJ/mol
3. The forward reaction of a reversible reaction is exothermic. – The backward reaction is endothermic.

   The forward reaction of a reversible reaction is endothermic. – The backward reaction is exothermic.
4. Colour change of the powder from blue to white.
5. a) A reversible reaction and closed system.
   b) They are the same.
   c) They remain constant.
6. Equilibrium is reached quicker / Increases rate of forward and reverse reaction by the same amount.

Page 89

1. To predict the effect on position of equilibrium / yield of product when conditions of an equilibrium reaction are changed.
2. For a reversible reaction at equilibrium, the system will oppose any change to the conditions.

3. decreases; increases; decreases; increases
4. a) It shifts to favour endothermic reaction / increases rate of endothermic reaction.
   b) It shifts to favour exothermic reaction / increases rate of exothermic reaction.
5. a) It shifts to the side with the lowest amount of gas.
   b) It shifts to the side with the highest amount of gas.
   c) No effect.

Page 91

1. a) Ancient biomass / plankton
   b) In rocks
2. A mixture of hydrocarbons with similar boiling points.
3. Evaporation and condensation
4. Hydrogen; Carbon
5. The larger the molecule, the less flammable it is.
6. The fuel is fully oxidised.
7. Single covalent bonds
8. hydrocarbon + oxygen → carbon dioxide + water

Page 93

1. The endothermic / thermal decomposition of long-chain hydrocarbons to make shorter, more useful, hydrocarbons.
2. Alkanes and alkenes / shorter chain hydrocarbons
3. It is vapourised, passed over a catalyst and undergoes thermal decomposition.
4. a) Catalyst (zeolite / aluminium oxide / silicon dioxide) and temperatures of 550°C.
   b) Petrol
5. a) Steam, high pressure and temperatures of more than 800°C.
   b) Polymers
6. a) They only contain hydrogen and carbon.
   b) They contain two fewer hydrogen atoms than the alkane with the same number of carbon atoms. They contain double bonds.
   c) C = C, carbon carbon double bond
7. Bromine water
8. C_3H_6

Page 95

1. a) Carbon dioxide and water
   b) Carbon dioxide, carbon (soot), carbon monoxide and water
2. Addition reaction
3. Butanol

4. **a)** Carbon dioxide and water
   b) Carbon dioxide, carbon (soot), carbon monoxide and water
5. -OH
6. Salt and hydrogen
7. -COOH
8. It partially ionises in solution.
9. Ester and water

Page 97

1. **a)** A long-chain molecule made of many repeating units.
   b) A small molecule that can join to other small molecules to make a polymer.
   c) The chemical reaction to make a polymer.
2. Alkenes
3. 1
4. Only one product / All atoms from the reactants are used in the desirable product.
5. Butene
6. Polypentene
7. 2
8. A polymer and a small molecule, e.g. water.
9. Diol and dicarboxylic acid

Page 99

1. Glucose
2. Glucose
3. 2
4.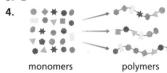

   monomers   polymers
5. Polypeptide / protein
6. In the nucleus of eukaryotic cells. (*Accept* In viruses.)
7. To encode genetic information.
8. Nucleotides
9. Double helix

Page 101

1. Nothing added / in its natural state.
2. Only one substance.
3. Water molecules only.
4. It contains more than one substance and is therefore a mixture.
5. A single type of element or a single type of compound.
6. **a)** A mixture designed for a specific purpose.
   b) It contains more than one substance.
7. *Any one of:* fuels; cleaning agents; paints; medicines; alloys; foods; fertilisers
8. It contains more than one substance.
9. For a particular function / purpose in the end product.

Page 103

1. **a)-b)**
2. The different parts of the mixture are attracted to the paper and solvent by different amounts and this causes them to separate.
3. The results of a chromatography experiment.
4. pure; impure; mixture
5. Retention factor / the ratio of the distance moved by a compound (centre of spot from origin) to the distance moved by the solvent.
6. The line that marks where the solvent travelled to in the paper / stationary phase.
7. **a)** R_f value $= \dfrac{\text{distance travelled by substance}}{\text{distance travelled by solvent}}$
   b) The identity of a substance.

Page 105

1. Seeing bubbles and hearing fizzing
2. One of the products is a gas and is lost to the atmosphere.
3. Displacement of water – When the gas has low solubility in water; Downward delivery – When the gas is denser than air; Upward delivery – When the gas is less dense than air
4. **a)** Hydrogen
   b) Carbon dioxide (*Accept* chlorine)
   c) Hydrogen
5. Lighted splint and hear a squeaky pop.
6. It relights.
7. Limewater
8. Litmus
9. It decolourises / bleaches / turns white.

Page 107

1. Some metal ions produce a characteristic flame colour.
2. Yellow / orange flame – Sodium; Lilac flame – Potassium; Brick red flame – Calcium
3. An insoluble solid made from a reaction between two solutions
4. White

5. Both form white precipitates, but the aluminium precipitate dissolves in excess sodium hydroxide whilst the magnesium one doesn't.
6. **a)** Cu^{2+} **b)** green
7. $CuSO_4(aq) + 2NaOH(aq) \rightarrow Na_2SO_4(aq) + Cu(OH)_2(s)$

Page 109
1. CO_3^{2-}
2. Effervescence / bubbles and fizzing
3. It changes from colourless to cloudy.
4. Negative ion
5. Nitric acid and silver nitrate
6. White
7. $Br^-(aq) + Ag^+(aq) \rightarrow AgBr(s)$
8. Iodide
9. Hydrochloric acid and barium chloride
10. $Ba^{2+}(aq) + SO_4^{2-}(aq) \rightarrow BaSO_4(s)$

Page 111
1. Using a machine to analyse elements and compounds.
2. Due to developments in technology.
3. **a)** Rapid, more sensitive and more accurate.
   b) More expensive and need specialist training.
4. Metal ions
5. Spectra
6. The measuring instrument which measures the wavelengths of light given out by the flame.
7. Flame tests
8. **a)** Mixtures can be analysed and more information can be given (more ions can be identified and concentrations measured).
   b) Cheaper and easier to carry out.

Page 113
1. The envelope of gas around our planet.
2. **a)** Nitrogen **b)** 20% **c)** Four fifths
3. Mars and Venus
4. Volcanic activity
5. Algae
6. Photosynthesis
7. Sedimentary
8. *Any three from:* dissolved into oceans; formed sedimentary rocks; formed fossil fuels; used in photosynthesis

Page 115
1. Water vapour, carbon dioxide and methane
2. To keep global temperatures high enough to support life.

3. They are increasing the proportion of gases, through combustion of fossil fuels, animal farming, rice paddies and landfill.
4. The increase in average world temperatures.
5. The long-term change in weather patterns and temperatures.
6. Increase in world temperatures cause the ice caps to melt and therefore sea level rises.
7. The total amount of greenhouse gases over the full life cycle of a product, service or event.
8. Carbon dioxide equivalent/CO_2e

Page 117
1. Sulfur
2. From engines with high temperature and pressure.
3. **a)** CO
   b) Incomplete combustion
   c) It reduces the oxygen-carrying capacity of the blood.
4. Sulfuric acid
5. Nitric acid
6. Rain with a pH lower than natural rain (5.5).
7. Carbon (soot) / solid particles
8. particles; reflect

Page 119
1. To provide warmth, shelter, food and transport.
2. **a)** A resource that is used chemically unchanged to support life and meet people's needs.
   b) A resource that has been chemically changed to make a new material.
3. **a)** Natural resource
   b) Synthetic resource
4. A resource that can be replaced as it is being used.
5. A resource that is being used up faster than the Earth can replace it.
6. Biomass – Renewable; Nuclear – Finite; Coal – Finite; Geothermal – Renewable; Wind – Renewable; Oil – Finite
7. Ensures the needs of the people are met today, whilst ensuring that there are enough resources for future generations too.

Page 121
1. Pure water is water that only contains water molecules. Potable water is any water that is safe to drink (including water that has safe, dissolved substances in).

2. Naturally occurring non-salty water, e.g. rain water. It can be found in rivers, ground water and lakes.
3. Filtering and sterilising
4. The water is passed through filter beds.
5. Ozone or UV light
6. When fresh water supplies are limited.
7. A lot of energy is used.
8. Reverse osmosis and distillation

Page 123
1. To prevent pollution.
2. Organic matter and microbes
3. Harmful chemicals and organic matter
4. Large, insoluble particles
5. a) The liquid part of sewage
   b) The solid part of sewage
   c) By sedimentation
6. Anaerobic bacteria
7. Aerobic bacteria
8. Bacteria

Page 125
1. Finite natural resource
2. Rocks that have a low percentage of metal in them.
3. *Any one of:* cooking pans; water pipes; electrical wires
4. a) Phytomining
   b) Bioleaching
5. A solution that contains metal ions.
6. a) *Any one of:* it preserves finite metal ore reserves; it can be used to clean up contaminated soils.
   b) It is time-consuming.
7. A more reactive metal takes the place of a less reactive metal in its compound.
8. a) They are reduced. (*Accept* gain electrons or become neutral atoms)
   b) They are reduced. (*Accept* gain electrons or become neutral atoms)

Page 127
1. The environmental impact of a product through all of its stages of manufacture, use and disposal.
2. a) Value judgements, numerical values of pollution effects
   b) Use of water, resources, energy sources and production of some wastes
   c) Facts without opinion / bias.
3. It is recycled or put in landfill.

4. A considered decision with sensible conclusions.
5. It must be cleaned and sterilised.
6. It is melted and re-cast / reformed.

Page 129
1. They oxidise
2. It changes the properties of the material.
3. a) Iron / Fe
   b) Hydrated iron(III) oxide
   c) Iron, air / oxygen and water
   d) $Fe_2O_3.xH_2O$
4. It prevents air / oxygen and water from reaching the material.
5. *Any one of:* grease; paint; plastic; metals
6. A more reactive metal corrodes instead of the metal material.
7. It is coated in an aluminium oxide layer which prevents air / oxygen and water from reaching the aluminium atoms.

Page 131
1. An alloy contains more than one substance.
2. Copper and Tin
3. Copper and Zinc
4. It is too soft.
5. 75%
6. Gold, silver, copper and zinc (*Accept* Au, Ag, Cu and Zn)
7. Strength
8. Low-carbon steel
9. Stainless steel
10. Chromium and nickel (*Accept* Cr and Ni)

Page 133
1. By shaping wet clay and heating in a furnace.
2. Sand, sodium carbonate and limestone
3. It has a high melting point. (*Accept* chemical resistant)
4. A long-chain molecule made of small repeating units.
5. They can be melted and easily reshaped.
6. a) thermosetting; thermosoftening
   b) thermosoftening; thermosetting
7. Matrix (*Accept* binder) and reinforcement
8. Polymer resin
9. It is stronger than each material on its own.

Page 135
1. Ammonia
2. Methane, Air, Water
3. Hydrogen and nitrogen

4. Iron-based
5. 200 atmospheres and 450°C
6. $N_2 + 3H_2 \rightleftharpoons 2NH_3$
7. Nitrogen, phosphorus and potassium
8. Titration and crystallisation
9. a) Mined
   b) Mined and treated with acid.

Pages 136–141 Mixed Questions

Pages 136–138

1. Halogens
2.

3. a) covalent
   b) 2,8,8
   c) Electrostatic force of attraction between oppositely charged ions.
4. When melting hydrogen chloride, no bonds are broken. There are only weak forces of attraction between molecules, but when sodium chloride is melted, many strong bonds must be broken. It requires more energy to break bonds than overcome forces of attraction.
5.

6. a) $(2 \times 14) + (4 \times 1) + (3 \times 16) = 80$
   b) $\frac{(2 \times 14)}{80} \times 100 = 35\%$
7. a) Oxidation is gain of oxygen (*Accept* loss of electrons).
   b) magnesium + oxygen → magnesium oxide
8. a) Magnesium
   b) Copper sulfate
   c) copper sulfate + magnesium → magnesium sulfate + copper
9.

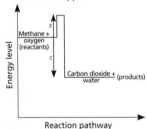

(Reactants higher in energy than products.

Activation energy marked (x), energy given out marked (z).)

10. *Any 6 of:* alkanes and alkenes are both hydrocarbons; both contain covalent bonds; both have intermolecular forces of attraction between the molecules; alkanes are saturated / only have single covalent bonds; alkenes are unsaturated / contain at least one C=C; alkanes have no reaction with bromine water but alkenes decolourise bromine water; both are examples of homologous series; general formula for alkanes is C_nH_{2n+2}; general formula for alkenes is C_nH_{2n}.

Pages 139–141

1. a) $N_2 + 2H_2 \rightleftharpoons 2NH_3$
   b) It is cooled and condenses into a liquid.
2. The minimum energy needed to start a reaction.
3. a) Concentration of acid
   b) g
   c) In a higher concentration there will be more acid particles in a given volume. This means that there will be more collisions in the same amount of time, so more successful collisions in a given time and therefore a faster rate of reaction.
4. a) Ethane
   b) C_nH_{2n+2}
5. a) Ca^{2+}
   b) The white precipitate formed doesn't dissolve with excess sodium hydroxide solution.
   c) I^-
6. a) Nitrogen
   b) O_2
7. 200 million years
8. Soot / carbon
9. a) Iron
   b) Oxygen is added to the iron. (*Accept* iron loses electrons)
   c) Galvanising is coating the steel object in zinc. This prevents the oxygen and water getting to the iron / barrier protection. Zinc is more reactive than iron so zinc will corrode / oxidise instead of iron in the steel / sacrificial protection.
10. Fractional distillation
11. a) A molecule that contains only hydrogen and carbon
    b) Cracking